HOT TOPICS

GENETIC ENGINEERING
AND GENETICALLY MODIFIED ORGANISMS

By Michelle Denton

Portions of this book originally appeared in *Genetic Engineering* by Tina Kafka.

LUCENT
PRESS

Published in 2019 by
Lucent Press, an Imprint of Greenhaven Publishing, LLC
353 3rd Avenue
Suite 255
New York, NY 10010

Designer: Deanna Paternostro
Editor: Jennifer Lombardo

Library of Congress Cataloging-in-Publication Data

Names: Denton, Michelle, author.
Title: Genetic engineering and genetically modified organisms / Michelle Denton.
Description: New York : Lucent Press, [2019] | Series: Hot topics | Includes bibliographical references and index.
Identifiers: LCCN 2018030680 (print) | LCCN 2018032120 (ebook) | ISBN 9781534565043 (eBook) | ISBN 9781534565036 (library bound book) | ISBN 9781534565029 (paperback)
Subjects: LCSH: Genetic engineering. | Transgenic organisms.
Classification: LCC QH442 (ebook) | LCC QH442 .D46 2019 (print) | DDC 576.5–dc23
LC record available at https://lccn.loc.gov/2018030680

Printed in the United States of America

CPSIA compliance information: Batch #BW19KL: For further information contact Greenhaven Publishing LLC, New York, New York at 1-844-317-7404.

Please visit our website, www.greenhavenpublishing.com. For a free color catalog of all our high-quality books, call toll free 1-844-317-7404 or fax 1-844-317-7405.

PICTURE CREDITS

Cover Himanshuroy Fulwala/EyeEm/Getty Images, p. 7 Miriam Doerr Martin Frommherz/Shutterstock.com, p. 10 Corbis/Corbis via Getty Images; p. 13 bernatets photo/Shutterstock.com; p. 16 ZSSD/Minden Pictures/Minden Pictures/Getty Images; p. 20 John Zich/The LIFE Images Collection/Getty Images; p. 22 Chelsea Lauren/FilmMagic/Getty Images; p. 25 Courtesy of Statista.com; p. 28 Edwin Remsburg/VW Pics via Getty Images; p. 31 Judd Christie Photography/Moment/Getty Images; p. 35 Roma Black/Shutterstock.com; p. 38 Igor Smirnoff/Shutterstock.com; p. 41 Aaron Haupt/Science Source/Getty Images; p. 44 PATRICK HERTZOG/AFP/Getty Images; p. 48 David Tadevosian/Shutterstock.com; p. 51 Hulton Archive/Getty Images; p. 55 Yann Guichaoua-Photos/Moment Unreleased/Getty Images; p. 59 photo-denver/Shutterstock.com; p. 64 Mitch Kezar/Design Pics/Getty Images; p. 66 FotoRequest/Shutterstock.com; p. 68 S.J. Krasemann/Photodisc/Getty Images; p. 71 Barcroft Media/Barcroft Media via Getty Images; p. 72 Dmitry Abaza/Shutterstock.com; p. 76 Yann hubert/Shutterstock.com; p. 81 MANJUNATH KIRAN/AFP/Getty Images; p. 84 PA Images via Getty Images.

ABOUT THE AUTHOR

Tyler Stevenson is a professional in the health insurance industry. He graduated from the University at Buffalo with a degree in sociology and originally went into banking before making the switch to health insurance. He works in Buffalo, New York, and lives in the City of Tonawanda with his wife; his two young daughters, Autumn and Summer; his dog; his two cats; and the family guinea pig, Pig-Pig.

CONTENTS

FOREWORD

Adolescence is a time when many people begin to take notice of the world around them. News channels, blogs, and talk radio shows are constantly promoting one view or another; very few are unbiased. Young people also hear conflicting information from parents, friends, teachers, and acquaintances. Often, they will hear only one side of an issue or be given flawed information. People who are trying to support a particular viewpoint may cite inaccurate facts and statistics on their blogs, and news programs present many conflicting views of important issues in our society. In a world where it seems everyone has a platform to share their thoughts, it can be difficult to find unbiased, accurate information about important issues.

It is not only facts that are important. In blog posts, in comments on online videos, and on talk shows, people will share opinions that are not necessarily true or false, but can still have a strong impact. For example, many young people struggle with their body image. Seeing or hearing negative comments about particular body types online can have a huge effect on the way someone views themselves and may lead to depression and anxiety. Although it is important not to keep information hidden from young people under the guise of protecting them, it is equally important to offer encouragement on issues that affect their mental health.

The titles in the Hot Topics series provide readers with different viewpoints on important issues in today's society. Many of these issues, such as domestic abuse and free speech rights online, are of immediate concern to young people. This series aims to give readers factual context on these crucial topics in a way that lets them form their own opinions. The facts presented throughout also serve to empower readers to help themselves or support people they know who are struggling with many of the challenges adolescents face today. Although negative viewpoints are not ignored or downplayed, this series allows young people to see that the challenges they face are not insurmountable. Abusive relationships can be ended, the internet can be navigated safely, and the environment can be saved.

Quotes encompassing all viewpoints are presented and cited so readers can trace them back to their original source, verifying for themselves whether the information comes from a reputable place. Additional books and websites are listed, giving readers a starting point from which to continue their own research. Chapter questions encourage discussion, allowing young people to hear and understand their classmates' points of view as they further solidify their own. Full-color photographs and enlightening charts provide a deeper understanding of the topics at hand. All of these features augment the informative text, helping young people understand the world they live in and formulate their own opinions concerning the best way they can improve it.

The Future Is Here

As society heads further into the 21st century, the impact of scientific advancement can be seen in every part of everyday life. People carry phones in their pockets that are more powerful than the first computers, which took up entire rooms. Humans interact with humanlike artificial intelligences such as Cortana, Alexa, and Siri almost daily, and these intelligences help people manage their lives. People watch rockets take cars into space, and the day humanity sets foot on Mars draws closer with every discovery and invention for space travel.

Some scientific advancement still scares people, however, in the same way it scared people when Nicolaus Copernicus and, later, Galileo Galilei claimed the sun, not Earth, was the center of the solar system. When Galileo observed the movements of the planets in greater detail through a telescope, it became clear to him that Earth was moving around the sun. His theories were rejected by the Catholic Church, however, and he was exiled from his home for heresy, or statements that went against the church's teachings. It was not until many years later that this idea was accepted as reality.

Today, genetic engineering is somewhat the same. Although it is surprisingly commonplace in modern times—from tomatoes spliced with fish genes to better resist the cold to the wide variety of dog breeds that have each been bred for certain traits—many people are nervous about what the future of genetic manipulation will bring and what it will mean for society. The ethics of humans tampering with nature raise moral and philosophical questions that both scientists and the public are struggling with. Where is the line, and what happens if it is crossed?

The Progress of Science

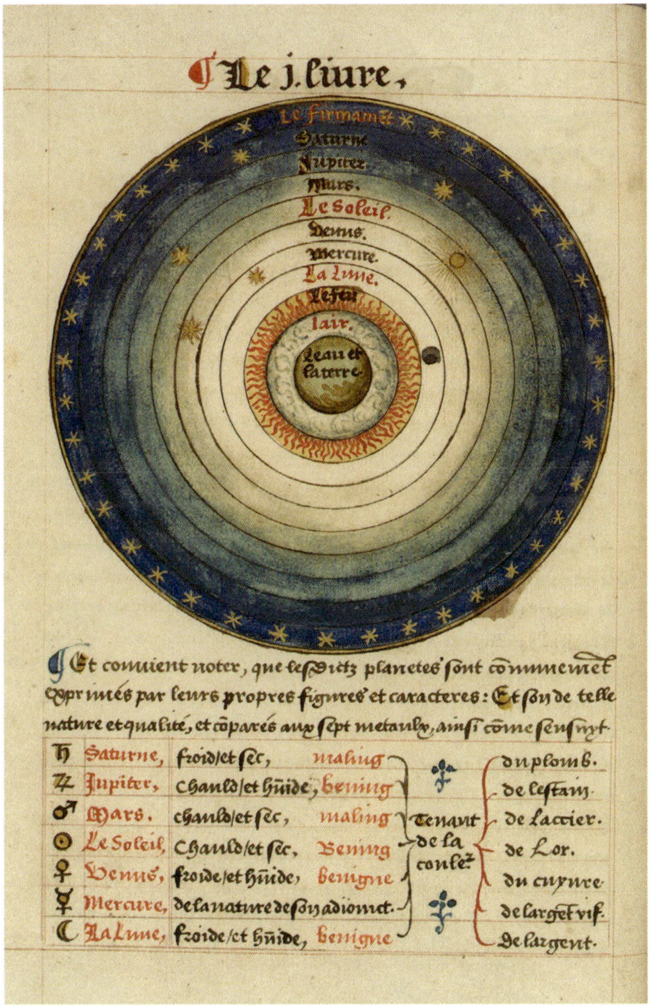

Shown here is the Earth-centric model of the universe that Copernicus and Galileo disproved. Just as their astronomical theories were rejected out of fear, modern-day scientists meet similar resistance to their genetic discoveries.

Deep questions arise along with the rapidly growing science of genetic engineering. Scientists and others argue about how to draw the line between ethical and unethical research, whether genetic engineering research should be limited, and who has the final say on these issues. Scientists, politicians, and various interest groups debate the safety of genetically modified organisms (GMOs) and how much risk should be allowed in the search for solutions to urgent health problems. However, even as the debate rages, biologists continue to move forward and look more closely at the most basic mechanisms of life. As they continue to fine-tune their focus, the power of genetic engineers to manipulate those mechanisms grows mightier.

The foundation of genetic engineering, like that of all science, is built upon past achievements. In the 1950s,

Rosalind Franklin discovered the double helix structure of the deoxyribonucleic acid (DNA) molecule, which contains coded instructions that direct the structure and function of every part of every living organism. Her experiments revealed that the DNA molecule is shaped like a twisted ladder. Each rung of that ladder is a joined pair of chemicals called bases. The pairs are composed of either adenine and thymine (A-T or T-A), or guanine and cytosine (G-C or C-G). Every living organism—plant or animal—contains these same four bases arranged in pairs along strands of sugar and phosphate. A gene is a specific sequence, or pattern, of these chemical pairs. Differences between individuals and even between species lie solely in the sequence of these chemicals along the strand.

Unfortunately, Franklin's coworkers at King's College in London, England, showed her work to scientists James Watson and Francis Crick, who took credit for the discovery. However, many view the work of two scientists in the 1970s, Herb Boyer and Stanley Cohen, as the most direct link between past research and modern genetic engineering. Boyer and Cohen found a way to cut DNA apart and recombine the parts in new ways. The general term for the resulting molecule is recombinant DNA.

Shortly thereafter, scientists found they could transfer genes from animals and plants into bacteria. The bacteria, in turn, would then work like tiny factories, churning out large quantities of the hormones or other chemicals directed by the transplanted gene. In 2004, Boyer and Cohen

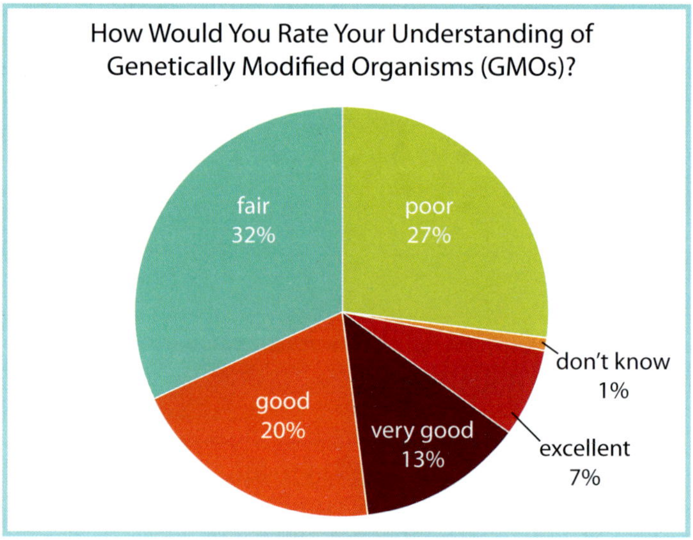

Genetic engineering is a complex topic, and many people do not fully understand it, as this information from the Annenberg Public Policy Center shows.

won the prestigious Albany Medical Center Prize in Medicine and Biomedical Research for this research that set the stage for modern genetic engineering.

It is frequently difficult to gauge the importance of a scientific advance at the time it is taking place. No one predicted, for example, that the discovery of the structure of the DNA molecule would change the criminal justice system by allowing police to compare a suspect's DNA with samples found at a crime scene. Nor did computer scientists foresee in the 1970s that computers would shrink dramatically in size, that people would create social networks online, or that vast amounts of information would be available at no cost to anyone with access to a computer small enough to fit on a person's lap.

In the same way, the applications of genetic engineering are just beginning to reveal themselves. The possibilities of recombining DNA from different species and directing organisms to produce medicines, chemicals, organs for human transplant, and new foods appear as limitless as the human imagination. Scientists are learning to identify sequences along the DNA molecule that are responsible for certain diseases such as diabetes, Alzheimer's disease, Parkinson's disease, cancer, and heart disease, among others. They are also searching for key sequences that might be responsible for outstanding qualities such as musical talent or athletic ability. Once these sequences are identified, scientists could gain the ability to add or subtract them from a person's DNA.

Old Science, New Name

Though the science of genetics is relatively new, even the first farmers performed their own version of it. They learned to save the seeds each year from the fattest barley, the tastiest corn, and the most productive wheat. They sowed those seeds the following year in the hope of improving their harvest.

Over time, crops changed to reflect the tastes of the farmers and those they fed. Domesticated animals followed a similar path. The tastiest hogs, the most productive cows, and the best chickens, among others, were bred selectively to improve the stock. Robert Krulwich, a science correspondent for National

Public Radio (NPR), explained on a segment for *Morning Edition* in 2008 how genetic engineering has changed over time: "Humans have been doing this for centuries, of course, using the breeding techniques that have produced Chihuahuas and poodles, and different varieties of apples and bananas, but bio-engineering, tinkering with the DNA, produces those changes much more quickly."[1]

An article in the *New York Times* published in June 1987 illustrates Krulwich's point. The article outlined some possible future scenarios in the rising field of modern genetic engineering. At the time, these ideas seemed like science fiction: plants that could produce their own pesticides and weed killers, the ability to select the sex of animals by manipulating the cells that produce eggs and sperm, the transfer of genes from one species to another, the birth of cloned mammals, and the ability of an animal of one species to gestate (grow in the womb) and give birth to an animal of a different species.

Not only have these scenarios happened, some now seem almost commonplace. In the 1990s, corn and cotton were genetically modified to produce their own pesticides. In 1996, Dolly, a sheep cloned from the udder cell of another sheep, was born in Scotland. In 2001, Bessie the cow gave birth to Noah, an Asian guar, which is an endangered breed of an ox. However, the gaur died two days after it was born.

The successful cloning of Dolly the sheep (shown here) showed the world that genetically engineering animals was possible.

Scientists have also genetically modified rice to contain high amounts of vitamin A by transferring daffodil and bacteria genes into the rice's genome, or genetic makeup. Animals, including dogs, cats, and even a bull, have been cloned. Pigs implanted with human genes grow organs for human transplant. The gene responsible for spiderweb silk has been engineered into goats.

Genetic engineering is already creating new paths in more directions—and more quickly—than anyone ever imagined. Bioengineers are developing computer programs that read the sequence of an individual's genome quickly. In 2003, the first full human genome sequence was completed. The Human Genome Project, as it was known, took 13 years and cost $437 million to complete. By 2008, the time needed to read a human genome was reduced to four weeks and cost $100,000. Some scientists predicted at the time that by 2015, new parents would pay $10 to have their baby's DNA sequence read at birth and placed in their medical file. This turned out to be an overly ambitious prediction, although it has become increasingly easy to test babies for certain diseases and disorders long before they are born—a practice that is not without controversy.

The knowledge that entire organisms are just various arrangements of a limited number of chemicals has fueled an entirely new field of genetic engineering called synthetic biology, or synbio. This process is, in effect, the reverse of reading a genome's sequence. Instead of reading an existing genome, scientists use the basic ingredients to build an entirely new organism from scratch. This, of course, raises many new questions. Is a life form just a machine made of parts strung together, each of which has a specific function? Some bioengineers think so. This is another issue society must struggle with as the science of genetic engineering moves steadily forward.

CHAPTER 1

Farming Pharmaceuticals

Plants and other living things have been used to provide medicine for thousands of years; for example, morphine, a strong painkiller, is produced from a common wildflower called the opium poppy. Penicillin, which is used to treat many infections, is the natural product of a mold. Some of these substances are now made in a lab rather than gathered from actual plants, but their chemical structure is basically the same.

Though medicines made from plants and even animal parts are not new, each species of medicinal plant or animal part has been kept mostly separate until the past few decades. Medicine might have combinations of different natural products, but they each came from a separate source. Today, however, the practice of "pharming," or genetically modifying plants and animals to produce pharmaceuticals, crosses the food people eat with the medication they need. A single organism, such as a bacterium, a pig, or a stalk of corn, can create and carry medical compounds they never could naturally.

Once genetic engineers learned to cut and recombine genes, the possibilities of developing drugs to treat human diseases and even creating organs for human transplant became almost limitless. Genetic engineering is still a new field, but already, the high

Even today, people rely on the natural properties of plants to make medicine.

number of possible medical applications of this new technology is apparent. Freeman J. Dyson, formerly a physicist on the faculty of the Institute for Advanced Studies in Princeton, New Jersey, speaks and writes frequently about the future of new technology. In his book *The Sun, the Genome, and the Internet*, he referred to the future of genetic engineering when he wrote, "The medical scientists of today will be leading revolutions into directions that we can only dimly discern."[2] Dyson is not alone in expressing awe at what may be possible when drug therapies are no longer limited by the genomes of individual species.

BETTER FOR EVERYONE

"Genetically modified animals should be a boon [advantage] both for consumers, who may ultimately gain healthier foods and access to scarce medications, and for agricultural producers, who may cut costs with disease-resistant or faster-growing animals."

–*New York Times* editorial

Sheep and Goats Helping Humans

Shortly after the celebrated birth of Dolly the cloned sheep, Sir Ian Wilmut, the scientist who cloned Dolly, used a modified version of the process to clone Polly, another sheep. This time, however, Wilmut added a human gene to the sheep cell during the cloning process. His goal was to develop a treatment for hemophilia, a genetic disorder in which people lack the protein that helps blood thicken and clot after an injury. People who have hemophilia risk bleeding to death, even from minor bumps and bruises.

When Polly matured, her milk contained the lifesaving human clotting protein missing in hemophiliacs. Scientists hoped hemophiliacs who drank the milk from the genetically engineered (GE) sheep would then be able to form blood clots, and Wilmut envisioned creating herds of GE sheep to treat this disorder. Unfortunately, this did not end up happening; drinking the milk had no effect on the disease, proving that the path of scientific progress is not always smooth. Many experiments fail before one succeeds. However, the fact that Polly's milk did contain the clotting protein showed that the biotech industry—propelled by advances in genetic engineering—was taking flight.

While the Polly experiment failed, another one succeeded. Thrombosis, a condition in which a blood clot forms and blocks a blood vessel, is the cause of heart attacks and strokes. Some people are more likely to develop blood clots than others. A biotechnology company in Massachusetts, rEVO Biologics (formerly GTC Biotherapeutics), genetically engineered goats to develop a drug called ATryn to prevent thrombosis. ATryn is produced when "pharmers"—scientists who develop pharmaceuticals—engineer a human gene that produces antithrombin, the protein that prevents clotting, into newly fertilized goat embryos.

An article titled "Pharm Animals Crank Out Drugs," published on the *Wired* website, explained the purpose of the 1,400 goats that graze on rEVO Biologics' rolling Massachusetts farmland: "The animals are sophisticated drug incubators, with millions of dollars of potential profit accumulating in their udders each day."[3] When the GE female goats mature, their milk contains the protein antithrombin, which is then purified into a form that can be administered to humans.

In the past, tiny amounts of this protein could be gathered only from human blood donations, a process that is inefficient and expensive. In contrast, rEVO said in 2009, one goat produced as much antithrombin each year as 90,000 human blood donors.

Chickens That Lay Medical Eggs

Goats and sheep are not the only farm animals that have had human genes implanted in their DNA. For more than 25 years, Scottish biologist Helen Sang has been working on producing chicken eggs enhanced with proteins for treating a variety of human

In several separate studies, scientists have genetically engineered chickens in an attempt to produce disease-fighting eggs.

diseases. Since the whites, or albumin, of chicken eggs already contain high concentrations of protein, fortifying chicken eggs with human proteins seems like it would not be difficult. Chickens, like rabbits, reproduce quickly. Each chicken lays many eggs, which means many patients could potentially be treated. In 2007, a team of researchers led by Dr. Sang at the Roslin Institute in Scotland successfully engineered hens that laid eggs containing cancer-fighting proteins. However, since new medical treatments take a long time to test, there have been few updates on the progress of this project.

Scientists at Origen Therapeutics, a biotech firm in northern California, are also working with chickens to develop eggs that produce human proteins. Scientists there hope eventually to breed chickens with eggs capable of producing the entire range of human antibodies to treat many different diseases. Elizabeth Svoboda, a science writer for *Wired*, explained how this accomplishment might ease the process of producing medicines: "If the company succeeds, harvesting compounds for drug therapies will be a little like choosing a flavor from a soda fountain."[4] However, even if it works, this technology is still years away from becoming available.

Engineering Diseases in Animals

While many researchers work on enhancing the milk, blood, and eggs of animals to provide ready supplies of human medicines, other scientists perform genetic modifications to deliberately cause certain diseases in animals. In this way, scientists study exactly what goes wrong on the genetic level when humans develop the same diseases. For example, rhesus monkeys have been genetically engineered to develop Huntington's disease, a devastating progressive illness that results in uncontrolled tremors, memory loss, personality changes, and ultimately, death.

Animal rights activists, however, do not agree with using animals in this way to benefit humans. While acknowledging that medical research is important, Maggy Jennings, a spokesperson for the Royal Society of the Prevention of Cruelty to Animals (RSPCA) in England, said, "There are many different ways of carrying out research on these diseases without using primates. The animal suffering involved, in our view, is considerable."[5]

Anthony Chan, who led a study in the United Kingdom (UK) on the monkeys, argued that it would be unethical not to use every possible means to find a cure for illnesses such as Huntington's. The gene responsible for this disease was discovered in 1993, but so far, scientists have been unable to find a way to reverse the brain deterioration that results in the disease's symptoms. Huntington's disease is particularly problematic, as it generally does not become evident until midlife. By then, many victims of this disease have already had children. Each child of a Huntington's victim then has a 50 percent chance of developing the disease if they have inherited the gene responsible for the disease.

Glowing Pigs

Primates are not the only type of animal used in scientific studies. Scientists at National Taiwan University in 2006, for example, successfully transferred the fluorescent green protein from jellyfish into pig embryos. The internal organs of the resulting piglets then glowed green under ultraviolet light. Additionally, the piglets had a little bit of green around their ears, eyes, and snouts. The study team was excited when the glow trait was passed from one of the genetically altered pigs to two of its piglets.

In 2006 and 2013, scientists spliced jellyfish DNA into fertilized pig embryos, creating pigs that glow in the right conditions. Someday, this could have important medical benefits for humans.

This proved two points: The pig with the jellyfish gene could have piglets, and the altered gene could be inherited. This genetic engineering success could prove important to stem cell research and to the search for a way to monitor the health of healing tissue in organ transplants. Tissue that glows is easy to track under ultraviolet light.

Other researchers say the glow is not an important trait. In 2013, scientists at the South China Agriculture University also used jellyfish DNA to create glowing pigs, but the main goal was to see if the pigs would absorb the jellyfish DNA. According to bioscientist Dr. Stefan Moisyadi, "The green is only a marker to show that it's working easily."[6] These researchers hope that by proving one animal can absorb DNA from a different animal, this knowledge can be used to make cheaper, more efficient medical treatments. However, not everyone supports this research because the animals might still have to be killed—for instance, if a human heart for a transplant patient is grown in a pig's body.

Animal rights activists object to animals dying for the sake of human medical research. They also object to any animal experimentation at all. As new technology, such as computer programs that let researchers model how a disease progresses in the human body, becomes available, an outspoken group voices the view that any laboratory research on animals, including manipulating their genetic makeup, is cruel and unnecessary. In response to such criticism, many medical schools and universities have reduced or eliminated the use of live animals in research studies.

Others argue, however, that a computer simulation cannot substitute for the observation of live animals.

Alternatives to Animal Testing

One of the most common uses of genetic manipulation in medical research is to cause an animal to develop a certain disease so treatments can be tested on them. A lot of the treatments and cures that are currently available for various diseases have come from this type of testing. However, many people object to it for several reasons. One is that human and animal bodies are different, so what works on a disease in an animal may not work in a human, which means wasted time and money in pursuit of a treatment that ends up failing. Another is that people believe animal testing is cruel. Even many scientists do not enjoy experimenting on animals, but most say there is no better alternative when it comes to trying to treat diseases.

To address this problem, researcher Donald Ingber has put human cells on a microchip in such a way that they behave like real organs. He started with a lung, and it mimics a human lung so well that it has successfully been used to test things such as the toxicity of a cancer drug and the effects of breathing in air pollution. The results the lung-on-a-chip has given in studies are so much more accurate than those that come from animal tests that Ingber and his team have been working on making other organs.

Insulin-Producing Bacteria

Surprisingly, even the smallest organisms can be genetically engineered. GE bacteria have been an important part of medicine since 1977. Type I diabetes is an autoimmune disorder in which a person's body does not produce enough insulin—a hormone that regulates a person's blood sugar. Every time someone eats, their blood sugar rises, and insulin makes sure it does not go too high or fall too low. Without insulin, the blood sugar builds up in the cells so they cannot take in nutrients, which causes illness and eventually death if it remains uncontrolled. To control their blood sugar, people with Type I diabetes must give themselves daily injections of insulin.

Before 1977, insulin was taken from cows and pigs, purified, and distributed to people with diabetes. However, that year, researchers discovered that they could combine a rat insulin gene with a bacterium to produce insulin. The process was improved over time, and in 1982, pharmaceutical company Eli Lilly became the first company to sell FDA-approved, GE human insulin. This is still how the majority of insulin medication is produced today.

Curing with Corn

Animals can lay GE eggs, give GE milk, and grow GE organs, but genetically engineering animals is expensive and labor-intensive. Each animal needs space to live and requires expensive ongoing care. It is more practical and cost efficient to grow GE vegetables than it is to genetically engineer the DNA of animals.

While most companies genetically engineer food to maximize production and resist pesticides that can kill plants, others have tried to use genetic engineering to produce medications for people. In the early 2000s, a biotechnology company called Epicyte Pharmaceuticals implanted corn with a human gene that produces an antibody, or natural defense, against genital herpes, a sexually transmitted disease that causes painful sores around the genitals. There is no cure for herpes; the sores can be treated, but the virus will always live in the body. The company hoped the corn could be used to create a gel that would cure the disease when someone put it on their sores.

While growing GE vegetables and plants is a more cost-effective avenue than altering animals, it is also more difficult to maintain control over the resulting mutations.

However, the company experienced setbacks in its research, and in 2004, it had to close down.

Epicyte also became infamous in the early 2000s when it announced that it was genetically engineering corn to include anti-sperm antibodies. The company hoped the corn could be used to create a form of birth control. However, as with its herpes research, this research stalled and was not completed before the company folded.

Epicyte's research fueled people's fears about the potential misuse of genetic engineering. The gene that attacks sperm has become known as the Epicyte gene, and some people fear that it is still being secretly included in GE corn, since Epicyte sold its assets to other companies when it folded. Reports that the effect of the anti-sperm antibodies was not reversible have continued to fuel conspiracy theories that genetic engineering companies are trying to sterilize part or all of the human race. Researchers have found no proof of these claims, but the fear persists.

While some people believe companies are purposely using GE food for unethical purposes, others worry about mistakes. Unlike animals, in which breeding can be carefully monitored, seeds from crops spread with the breeze, which is difficult to control. Most companies promise not to grow GE crops near regular crops, but it is difficult to predict how far the wind or insects will carry pollen. Additionally, sometimes employees mix up GE food with regular food. This happened in 2000, when GE corn intended for animal feed was discovered in Taco Bell brand taco shells sold in grocery stores. Known as StarLink corn, the grain was genetically engineered to contain bacteria that made it toxic to the corn borer, a pest in the moth family that feeds on growing corn. In effect, the corn produces its own insecticide. Since the engineered corn was only approved for animals, not for human consumption, Kraft Foods was forced to recall millions of boxes of the taco shells. This incident is often used to highlight the dangers of cross-contamination between fields of growing crops.

Despite concerns about contamination of food crops, many biotechnology companies are moving forward to genetically

"Hypoallergenic" Kittens

Until 2015, $4,000 could buy a "no-sneeze" kitten from Allerca, a small California biotech company that claimed to have genetically engineered a breed of felines that would not cause allergic reactions in humans. People saw the Allerca cats as the ideal way to resolve the conflicts that arise in homes where cat lovers suffer from allergies. However, in 2013, journalists at ABC News began investigating the company's claims and found that many Allerca kittens had been given away because their owners were still having allergic reactions to them. Indoor Biotechnologies, an immunodiagnostics and biotechnology company, did tests on the cats' fur and discovered that they were no more hypoallergenic than any other cat breed. Although some Allerca customers were satisfied with their new kittens, the company closed in 2015.

A California biotech company claimed to have crafted a hypoallergenic cat, but tests showed that the expensive, genetically engineered kittens were no more "sneeze-free" than regular cats such as this one.

engineer corn and other grains. While many applaud this technology, others are concerned about it. Larry Bohlen, founder of NaturalCheck, which makes food testing kits to check for contamination and fake products, represents many who fear that "just one mistake by a biotech company and we'll be eating other people's prescription drugs in our corn flakes."[7]

Contaminating Crops

Many people believe biopharming—the process of genetically engineering plants and animals—could be a good thing. However, biopharming and transgenic organisms (plants or animals that contain the genes of other plants or animals) got a bad reputation early on because of mistakes and setbacks. One of the biggest scandals happened in 2002, when soybeans in Nebraska were contaminated by corn that had been genetically engineered to contain a vaccine for pig diarrhea. The soybeans were used in many products, including baby food and ice cream. ProdiGene, a leading biotech firm, was fined $250,000 and had to pay almost $3 million to compensate farmers for the destruction and quarantine of the crops. Additionally, many of the biopharming experiments ended up failing because of technical problems—for example, Epicyte's proposed herpes cure. People got so nervous about biopharming that many investors stopped funding it, so research stalled and many companies failed.

New regulations set by the United States Department of Agriculture (USDA) and the Food and Drug Administration (FDA) as well as improved technology gave new life to the biopharming industry around 2012, with new plants and animals being approved for drug production. For instance, in 2015, a research team from Stanford University found that etoposide, a component of chemotherapy medication that is used to treat some types of cancer, could be grown in tobacco plants. Currently, podophyllotoxin—a chemical that is necessary to create etoposide—can only be found in a rare, endangered Himalayan plant called the Mayapple plant. Splicing, or implanting, Mayapple plant genes into tobacco plants could make etoposide more readily available and prevent the Mayapple plant from going extinct. Research on this and other transgenic projects is ongoing.

Keeping GE Crops on the Pharm

Those who believe genetic engineering offers hope for both treatments and profits conflict on many issues with those who resist this new technology. Containment is perhaps the most commonly cited problem. Many critics point to the difficulty of containing GE crops to keep them out of the food supply. Farmers have tried various strategies, some more successful than others. One method used by some of the largest biotech firms involves growing test crops in sealed labs or in fields that are miles away from other crops. However, some environmental groups do not consider increasing the distance between crops a workable solution. A coalition of several environmental groups filed a lawsuit against the USDA in 2003, hoping to ban the use of food crops for pharmaceutical use and restrict the plants to greenhouses. At the time, François Arcand, president of the Conference on Plant-Made Pharmaceuticals held that year in Quebec, Canada, said, "Molecular farming represents the pharmaceutical industry's best opportunity to strike a serious blow against such global diseases as AIDS, Alzheimer's and cancer."[8] A spokesperson for Monsanto, one of the best-known biopharming companies—owned by the pharmaceutical company Bayer as of 2018 and no longer operating under the Monsanto name—agreed, warning that restricting biopharms to greenhouses would set the industry back by up to 20 years. Although test crops are grown in isolation, approved crops that have passed safety tests are grown on normal fields. Most American farmers buy Monsanto's seeds, which are genetically engineered for purposes such as an increase in the amount of food they grow and resistance to pesticides, to grow on their own fields.

In an attempt to avoid criticism and ensure the isolation of pharm crops, a researcher from Purdue University in Indiana keeps his pharm underground. Controlled Pharming Ventures promotes underground cave pharming in an attempt to control the unintended spread of stray pollen. Doug Ausenbaugh, the company's founder, adapted a former limestone quarry and underground warehouse in Marengo, Indiana, to grow pharmaceutical corn, tobacco, tomatoes, and other plants that have altered DNA in order to produce a medicinal compound or vaccine. In addition, the scientists are working on ways to recycle dead plants to provide energy to fuel the lighting system. Moreover, cave pharming may also work alongside the growing popularity of farming without pesticides, since common agricultural pests do not thrive underground.

However, critics point out that cave pharming is not enough to protect food crops from contamination. They remind the public that shipping mishaps can also be blamed for contamination. Whether GE crops are restricted to caves, greenhouses, islands, or isolated farms, most people admit that it is impossible to guarantee with perfect certainty that pharm crops will not find their way into food crops. However, the public, scientists, politicians, and interest groups on both sides of the issue agree that the technology has already advanced too far to turn back. The benefits and the dangers that accompany it pose a dilemma. Most of those involved agree that it is difficult, if not impossible, to reconcile two interests that are in direct conflict with each other: genetically engineering food crops for purposes other than food and guaranteeing that the food supply is safe from contamination.

Regulating the Biotech Industry

In the United States, three government agencies regulate different aspects of the biotech industry. The Animal and Plant Health Inspection Service (APHIS) division of the USDA requires notification that tests will take place, issues permits to test certain varieties of plants that may be toxic, and then removes plants from further oversight once the tests are complete. The FDA ensures that foods made from these plants are safe for humans and animals to eat. The Environmental Protection Agency (EPA) is responsible for the safety of pesticides. FDA publications have stated that strict safety testing guidelines and consultation services help ensure safety. In the 1990s, it created the Plant Biotechnology Consultation Program specifically to work with GE food companies. Although consultation is voluntary, the FDA has stated that "GE plant developers routinely participate in it before bringing a new GE plant to market."[9] In fact, it is in a company's best interest to work with the FDA because a plant that makes people sick would give the company a bad reputation and hurt its profits. The consultants look at the new food to determine whether it might make someone sick and whether it is as nutritious as it would be if it were bred naturally. If an issue is found, the FDA works with the company to fix it; if it cannot be fixed, the food is generally not sold. In one example, "a soybean enriched with a protein from a Brazil nut wasn't brought to market, even as animal feed, because tests showed that it might trigger a reaction for people with an allergy to those nuts."[10]

However, many consumer groups do not like that consultation is voluntary. They say the GE system has no central authority and too many loopholes. Although decisions by these agencies are not supposed to be motivated by politics or pressure, they are vulnerable to both. Many believe the bottom line is not public interest but the interests of the biotech companies, whose future profits are determined by decisions made by these agencies. Concerned scientists and members of the public insist that companies value profit more than consumer safety.

Pharmaceutical companies stand to profit when products are deemed effective and safe. Consumer groups insist that the effects of GMOs are still unknown. They fear that GMOs will destroy the food supply and the environment and that once released, they cannot be recalled. One thing is certain: While the growing field of pharming is full of controversy and passionate opinions, the technology of transferring genes back and forth between animals, humans, and plants to produce medicines continues to march steadily forward.

The lack of a central governing agency makes it easier for companies to profit from genetic engineering. Their pursuit of profit makes many people distrust GE companies.

CHAPTER 2

Genetically Engineered Foods

Using GE food crops to make medicines is highly controversial, and changing the nutritional composition of food itself is no different. Both supporters and critics of genetic engineering have passionate feelings about food, which makes even the basic composition of everyday foods a source of disagreement. It is a basic need for survival, as well as many people's livelihoods, so the future of food is important to everyone on the planet. Supporters say that modifying food to last longer, grow bigger, and produce more are all necessary to solving problems such as food shortages and widespread hunger. Critics, however, worry about the health risks associated with eating GM food and the impact GM crops can have on the environment.

The Flavr Savr

The first GE food hit the produce section of grocery stores in 1994. Calgene, a biotech company in California, began to market its Flavr Savr tomato. By transferring the information from a gene for antibiotic resistance into the tomato's genome, genetic engineers created a tomato that would not rot as quickly as a conventional tomato. This solved a problem that had long plagued tomato growers: In order to preserve tomatoes and keep them fresh, farmers had to pick them while they were green and then chemically ripen them after shipping. Not only was chemical ripening distasteful to the public, the gas-ripened tomatoes lacked flavor. The Flavr Savr tomato, on the other hand, was tougher than the ordinary tomato. Farmers could allow it to ripen on the vine and ship it, and

Tomatoes that were genetically engineered to stay fresh longer were introduced in 1994.

it would arrive in the supermarket still red, ripe, juicy, and ready to eat, rather than bruised and leaking juice.

To reassure the public that its product was safe, Calgene requested a consultation from the FDA, which ruled in favor of the Flavr Savr tomato. However, as hardy as it was, Calgene and its tomato did not last long. A 2000 report published in *California Agriculture*, a publication of the University of California's Division of Agriculture and Natural Resources, stated that the decline in sales started in the UK. The tomato had been doing well there until 1998, when Dr. Arpad Pusztai appeared on a British television broadcast and "announced his conclusion that feeding rats genetically modified potatoes resulted in biological effects that 'could' be attributed to the process of genetic engineering, rather than to the product of the introduced gene."[11] Research into these claims found them to be false, but by then, the public had become nervous about GMOs, including the Flavr Savr tomato. Some FDA officials also raised questions about the tomato's safety, based on lab studies on rats. Eventually, Calgene sold a majority of its shares to Monsanto, which stopped producing the Flavr Savr tomato; the company focused on engineering plants to

The Terminator Gene

Named after *The Terminator*, a movie in which actor and former California governor Arnold Schwarzenegger plays a robotic killer, the terminator gene could potentially be genetically engineered into plant seeds. When these seeds ripened, molecular switches would automatically turn off the seed's ability to reproduce so the seed could no longer sprout. The terminator technology was first patented by Delta and Pine Land Company (D&PL), the world's largest cottonseed company, in 1998. D&PL was purchased by Monsanto, which was in turn purchased by Bayer, making Bayer the current owner of the patent (a document that allows them to be the only ones who sell their product). One major concern raised by opponents of GMOs is that implanting the terminator gene into any of the seeds the company sells could devastate half of the world's farmers; they say these farmers depend on the seeds from each year's harvest to plant the following year's crop. Once this gene is engineered into the seed they purchase, farmers must then purchase new seed for each planting season, an expense critics say they cannot afford.

However, Monsanto has never created products containing the terminator gene and has pledged not to because plants that cannot reproduce would negatively affect the world's food supply. Additionally, the belief that farmers depend on previous years' seeds is false. National Public Radio (NPR) reported, "By the time Monsanto got into the seed business, most farmers in the U.S. and Europe were already relying on seed that they bought every year from older seed companies."[1] Seed companies such as Monsanto require farmers to buy new GE seeds from them each year because that is how they stay in business; they created the new types of seeds, and they own the patent on them. Monsanto sues farmers who reuse their patented seeds, as it is a violation of the contract they sign when they agree to do business with Monsanto. Amanda, a lawyer whose family grows soybeans and corn, explained that farmers generally prefer to buy from seed companies–rather than save and reuse their seeds–for several reasons: It is easier, cheaper, and more efficient, and it guarantees that the genetic traits that help the plant sell well are present. Farmers who want to reuse their own seeds can choose to buy ones that are not patented, but most do not. On her blog, she wrote, "The bottom line is that we're making decisions on our farms that are right for us, the environment, and consumers. In general, saving seeds from one crop to use for the next is not in line with those goals."[2]

1. Dan Charles, "Top Five Myths of Genetically Modified Seeds, Busted," NPR, October 18, 2012. www.npr.org/sections/thesalt/2012/10/18/163034053/top-five-myths-of-genetically-modified-seeds-busted.

2. Amanda, "No, Farmers Don't Want to Save Seeds," *The Farmer's Daughter USA*, February 23, 2016. www.thefarmersdaughterusa.com/2016/02/no-farmers-dont-want-save-seeds.html.

produce more food rather than to resist bruising. Most tomatoes today are gas-ripened just as they were in the past, which has brought back people's dissatisfaction with their lack of flavor.

Fears About GMOs

Another GM tomato, which never made it to the market, was the Flounder tomato. The effort to create tomatoes that could withstand cold was spearheaded by DNA Plant Technology Corporation. Scientists there engineered a gene from flounder, a fish that lives in cold Arctic waters, into the DNA of tomatoes. The scientists hoped this gene transfer would create a tomato that could withstand frost in the field and resist cold damage in storage.

Though the experiment failed, the legend itself grew. Like many folk tales, this one has morphed, turning into a story of strawberries with fish genes. In reality, according to the Public Issues Education Project of Cornell University, "There are no published studies involving strawberries, no companies which have announced research or marketing plans for such a product, no government records of field testing such a plant, and no trace in the media to explain how this story may have originated."[12]

However, though these "fishberries" may be imaginary, the legend still flourishes. It is often used by critics to emphasize the foolishness of crossing natural genetic barriers, but these warnings display a misunderstanding about how genes work in GMOs. One article published on the website The Organic View asked, "How do you feel about the thought of crossing fish DNA with strawberries? I bet that would make an interesting topping for cheesecake."[13] However, inserting one or even several genes from a fish into a strawberry—or any food—would not make it taste at all like fish because DNA itself does not have a taste. The Genetic Literacy Project explained,

> *Inserting a fish gene into a tomato does not make a tomato "fishy" in the same way that humans cannot breathe underwater.*

> *We share some genes with fish,* but that doesn't make us fish. *Indeed, a single gene can hardly be considered "part of" a specific organism at all.*
>
> *Genes are not characteristic of organisms. An organism is characteristic of its combined genetic makeup.*[14]

Many people believe combining genes this way is unhealthy. Some fear that eating food with altered DNA could change the DNA of the person eating it. However, "genetic material doesn't get tacked on like pin the tail on the donkey. An 'added' gene isn't going to fall off and get stuck to yours. Bacteria-fighting enzymes and processes in your body are designed to prevent a genetic invasion."[15] Most people have been eating GMOs for years without noticing ill effects. For instance, most sugar in processed foods comes from sugar beets, and nearly all beets have been genetically altered to be bigger and last longer than they did in the past. The sugar that comes from these beets is identical to that from beets that have not been genetically engineered.

Others object on religious grounds; they dislike it because they believe people do not have the right to change nature, and many feel that these "unnatural" foods must be unhealthy or else they would not need to be made by humans. The Organic View wrote, "If nature wanted fishberries, they would have already evolved. Nature would have found a way to do what man has done. After all, it is not nice to mess with Mother Nature!"[16] However, genetic engineering has been taking place for years. Early farmers saved the best seeds from their crops and bred the animals that produced the most food to make sure only the best genes were passed on. This is what "turned small bunches of tiny kernels on tall grass 10,000 years ago into the big ears of juicy corn on the cob we have today."[17] Moving the process into a lab has not changed how it works; it has only sped up the results. Genetic engineering has allowed more food to be grown for people and animals in ways that benefit the environment. *Forbes* magazine explained,

Close to a billion people are malnourished [in the world], and almost 50 million in the United States are food insecure. By increasing corn yields, GMOs have tempered [reduced] the price spikes that have plagued grain markets causing food riots worldwide. GMOs also hold one of the keys to reducing the environmental impacts of agriculture. Since the introduction of GMOs, use of chemical herbicides has declined by 37 percent, and soil erosion has declined substantially.[18]

Some experts believe the world's population is now too large to feed without help from GMOs. However, others disagree. The Institute for Responsible Technology (IRT) issued a report in 2014 called *GMO Myths and Truths*, which stated that according to some genetic engineers, GMOs are not safe and are also not necessary to feed the world. One piece of evidence the report points to is that Europe still uses mostly non-GE crops but produces more food with less pesticide use than the United States. The report also noted that lab tests on mice as well as on isolated human cells have shown that some GMOs have toxic effects. People on both sides of the GMO issue often accuse the other side of covering up or ignoring important evidence, so it can be very difficult for a consumer to know what to believe.

GMOs in the Store

Although the first GE food to hit the grocery store aisles was a tomato, GE foods are actually far more common in other sections of the supermarket than they are in the produce section. Flour, oil, salad dressings, pie, chips, cookies, fried foods, and candy coatings often contain GM ingredients. Up to 80 percent of the processed food products sold in the United States contain GM ingredients that the FDA has said are safe for people to eat. The most commonly engineered foods, according to the *Huffington Post*, are corn, soy, squash, zucchini, alfalfa, canola, sugar beets, and milk. Some of these products, such as corn and soy, are used in products to sweeten and thicken foods. Roughly 81 percent of the corn in the United States is engineered to contain a gene that comes from a bacterium called *Bacillus thuringiensis*, known as Bt. Bt is toxic to certain insects, including the European corn borer, corn's most destructive enemy. When the Bt gene is inserted into corn DNA, the corn produces its own pesticide—bug killer is built right into the plant. Soybean farmers have also benefited from what is known as the first generation of GE foods. More than 90 percent of the U.S. soy crop is engineered with genes taken from bacteria that allow the plant to withstand sprayings of weed killers, such as Roundup—an herbicide often favored by soybean farmers. Soybean plants that have been genetically engineered in this way are often referred to as "Roundup Ready."

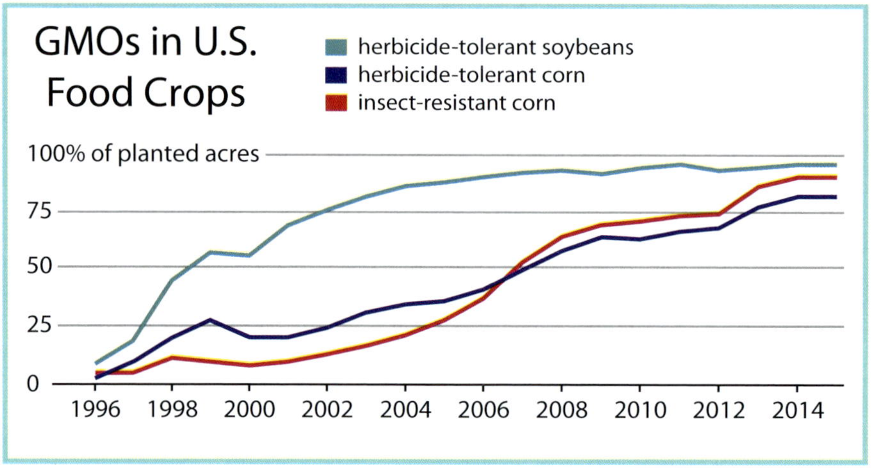

GMO use in the United States has been steadily rising, as this information from the Center for Agricultural Food and the Environment at the University of Massachusetts at Amherst shows. A study published in 2016 by the National Academies of Science, Engineering and Medicine found no strong evidence to suggest that GMOs are unsafe.

Critics believe these GE organisms threaten human health and the environment. They worry about GMO crops reducing plant diversity and killing too many insects, leading to the deaths of animals that eat those insects to survive. Supporters of genetic engineering believe the risks are minimal and the potential for solving some of the world's most persistent problems, such as hunger and malnourishment, is great.

The Next Step

Plants that produce their own bug weed killers are only the beginning. Biotech companies and university labs are working on more complex genetic engineering feats to make crops that are richer in nutrients such as vitamins, minerals, and omega-3 fatty acids. In some cases, scientists are working to turn genes off so that plants will lack proteins that cause allergies.

At Alabama Agricultural and Mechanical University, for example, scientist Hortense Dodo began working to develop hypoallergenic peanuts. She now owns a biotech company called IngateyGen and is working to develop the "Peanut4Life." Scientists elsewhere are developing tomatoes that are richer in folic acid and potatoes that contain high levels of calcium. At the University of California, Berkeley, Peggy Lemaux, a professor of plant and microbial biology, has worked to develop a variety of wheat that can be eaten by people with wheat allergies. However, cost has been a barrier to making it commercially available; Lemaux said that it would take between $10 million and $20 million to ensure the wheat is not a risk to the environment, humans, or other plants.

Supersized Fish

Scientists are also tinkering with the genes of foods other than grains in attempts to nourish the world's growing population. Research is well underway to genetically engineer both rainbow trout and salmon to grow bigger and meatier and to reach those sizes more quickly than conventional fish. The Agricultural Research Service, a government science agency, is working with private biotech companies to develop a variety of rainbow trout with an extra set of chromosomes. This extra set gives two benefits to the engineered trout. First, it ensures that the fish is sterile. These trout, therefore, cannot interbreed with wild trout, so there is no chance that they will pass on their modified genes. Also, the sterile trout grow bigger and faster, since energy that would otherwise be expended in breeding is devoted only to growing. However, there have been few updates on the progress of these fish in recent years.

DNA manipulation even extends to fish, which are being genetically engineered to grow bigger and more quickly in an effort to meet the demands of a hungry world.

On the other hand, the trout's saltwater cousin the salmon has been genetically engineered with much better results. AquaBounty Technologies in Massachusetts has gained FDA approval for a transgenic Atlantic salmon, known as the AquAdvantage salmon, that has been engineered with two genes. The first is a growth hormone gene from the Chinook salmon, and the second is a gene from another sea creature—the ocean pout—that keeps the growth hormone flowing year-round. Ordinarily, salmon grow primarily during the warm summer months. This GE salmon, however, reaches market weight in only 18 to 20 months, whereas conventional salmon takes 28 to 36 months to reach market weight. The AquAdvantage salmon is the first GE animal to be announced fit for human consumption by the FDA, but the decision has been widely controversial. In *TIME* magazine's report on the AquAdvantage salmon, it pointed out, "Most studies on animals that are fed genetically modified foods don't show serious health effects, though there are a few that hint at potential harm to organs like the kidneys, liver and heart, as well as increased risk of cancers and early death in these animals,"[19] which raises concerns about how humans' health will be affected by eating these fish.

Concerns About GE Animals and Animal Feed

The debate about GE animals is just as heated as arguments about GE crops. There are many different ways to engineer animals, but one of the most common is injecting cows with recombinant bovine growth hormone (rbGH)—also known as bovine somatotropin (bST)—which makes cows grow faster and produce more milk. The FDA has said meat and milk from these animals is safe for humans, and many people believe using rbGH has numerous benefits. Large herds of cows have a negative effect on the environment because land must be cleared for them to graze on; this reduces the number of trees that absorb carbon dioxide from the atmosphere, and the cows produce methane gas. Together, these two factors increase global warming. Producing more milk from fewer cows helps the environment as well as meeting consumer demand.

However, many people are unconvinced that milk from cows treated with rbGH is safe for humans to consume. These fears stem from reports that the increased amount of milk has caused an infection of cows' udders called mastitis, which means the cows must receive antibiotics. This has made people worry that they may be getting infected milk or that the antibiotics may be harmful to humans. Similar fears surround the food the animals are given. Some GE corn has been found to be suitable only for animal consumption, not for humans. This makes people wonder whether there are any health concerns that come from eating meat or drinking milk that comes from cows that have eaten the corn.

Additionally, there has been little research done on how rbGH may affect humans after they are exposed to it through the dairy products they consume, although studies comparing milk from cows with and without the added hormone have found few, if any, differences. Dale Baumett, the scientist who helped develop rbGH for Monsanto, thinks people's fears have no evidence behind them:

> *What I hear from farmers is this stuff is amazing. It's been looked at by not only FDA but the American Medical Association, American Academy of Pediatrics, American Dietetic Association, National Institute of Health. All the scientific groups. And there are areas where there is controversy in science, but the safety of milk from BST is not one of those areas.*[20]

Some farmers object to using the hormone for economic reasons; they say putting more milk on the market from large farms will put smaller farms out of business. Others believe it is bad for their cows, pointing to the increase in mastitis. This issue can be very confusing for the average consumer, as there is a lot of heated debate surrounding it. According to Gary Hirschberg of Stonyfield Farm, a yogurt company in New Hampshire, "Consumers are definitely confused and with good reason. The amount of propaganda around this issue on both sides has really been obscene [terrible]."[21] In the absence of a clear conclusion, several countries have banned rbGH either partially or totally. For instance, the European Union (EU) allows stores to sell imported dairy products from cows that have been treated with the hormone but has not allowed its own dairy farmers to use rbGH on their cows. People who want to avoid the issue should look for dairy products with a label explaining that the milk comes from cows that have not been treated with the hormone. This label may call the hormone by any of its several abbreviations, including rbGH or bST.

Grown in a Vat

Animal rights activists object to raising animals that will be killed for food, but one day, slaughtering an animal may not be a requirement for consuming a steak or a chicken nugget. Some scientists are working to bypass the animal altogether and grow meat in vats. Many science fiction authors have written about this since the early 1900s. One of the most recent was Margaret Atwood, the author of *The Handmaid's Tale*. In her 2003 novel *Oryx and Crake*, Atwood described a fictional food called ChickieNobs, which was created from chickens that were genetically engineered to grow only certain parts, such as breasts or drumsticks.

However, this idea has now gone beyond science fiction. The first to venture into this field of study was the late Morris Benjaminson, formerly a biology professor at Touro College in New York, who extracted goldfish muscle cells and chemically induced them to grow into fish fillets. He foresaw the day that chicken breasts would be mass-produced without the chickens and filet mignon produced without the steer. His research, however, was quickly brushed under the rug.

Since Benjaminson, many other people have attempted similar projects. In vitro culturing, a process which uses stem cells to create edible muscle tissue, has become popular among biotech startup companies. Some, such as New Wave Foods, have moved on to substitutes, such as algae, to mimic the texture of animal proteins, including shrimp. Jason Matheny, director of Intelligence Advanced Research Projects Activity (IARPA), envisions a future in which meat is grown in sheets of fat and muscle cells. The advantage, then, is that "In theory, with cultured meat, nothing is wasted, nothing suffers and nothing dies."[22] Matheny believes that consumers can be educated to accept vat-grown meat just as they now accept chickens and other livestock raised in what many consider less-than-ideal circumstances.

A PREDICTION FROM THE PAST

"Fifty years hence … we shall escape the absurdity of growing a whole chicken in order to eat the breast or wing, by growing these parts separately under a suitable medium."

–Winston Churchill, former prime minister of England, 1936

Meat without Harming Animals

If vegetarians avoid meat because they object to killing animals, vat-grown meat might increase the menu choices at vegetarian dinner tables. Theoretically, once scientists perfect the technology of culturing meat from stem cells and genetically engineering it to contain less fat and more protein, very few cells are needed.

Animals naturally shed thousands of cells each day, so critics might be hard-pressed to argue that vat meat exploits animals. In fact, to the surprise of many, the animal rights group People for the Ethical Treatment of Animals (PETA) announced in April 2008 that it would pay $1 million to the first person to develop a practical method of producing edible meat without an animal by 2012. The contest specified that the product must look and taste like real chicken. The chicken-like product would be prepared using a fried chicken recipe and would be judged by a panel of 10 PETA judges. The contest guidelines required that the company that submitted the winning entry also prove that its product was practical to make and sell. The winning company would be required to produce at least 2,000 pounds (970 kg) and distribute it to fast-food restaurants and grocery stores in 10 states over a period of 3 months. The cost also had to be close to that of conventional chicken. Ingrid Newkirk, a founder of PETA, explained that the contest itself was controversial within the organization, since many PETA members object to eating anything that even resembles meat, regardless of how it is produced.

The deadline for the contest was extended to 2014, but no one was able to claim the prize, so the offer expired. However, PETA still considered the contest a success because it caused the field of lab-grown meat to grow quickly. In 2017, PETA reported on what it called "cruelty-free" or "clean" meat from the company Memphis Meats, which is based in San Francisco, California. The company has created lab-grown beef, chicken, and duck that testers say taste like the real thing. In addition to being good for animals, lab-grown meat is good for the environment. Because the animal cells are grown in steel tanks, "Memphis Meats' method of meat production uses 1/10th the water and 1/100th the land that's currently used to raise billions of cows, pigs, and chickens each year. The process also produces less waste and cuts energy consumption in half."[23] Other companies, such as Mosa Meat and SuperMeat, are competing to be the first to sell lab-grown meat in stores. According to PETA, "The startups' current efforts are focused on lowering the cost of production, and slaughter-free 'clean meat' should hit the market [by 2022]."[24] Other companies are experimenting with making plant-based food taste, look, and smell like meat. The Impossible Burger, for instance, is made from potatoes, wheat, coconut oil, and other plants. However, the addition of a molecule called heme makes it taste like a beef burger.

A MYTH ABOUT MONSANTO

"A group of organic farmers ... recently sued Monsanto, asserting that GMOs might contaminate their crops and then Monsanto might accuse them of patent infringement. The farmers couldn't cite a single instance in which this had happened, though, and the judge dismissed the case."

–Dan Charles, food and agriculture correspondent for NPR

Dan Charles, "Top Five Myths of Genetically Modified Seeds, Busted," NPR, October 18, 2012, www.npr.org/sections/thesalt/2012/10/18/163034053/top-five-myths-of-genetically-modified-seeds-busted.

The Organic Food Movement

One by-product of the availability of GMOs—unintentional on the part of the biotechnology industry—is the growing organic food movement. In 2016, sales of organic foods reached more than $47 billion. That is still a small fraction of overall food sales, but the annual growth of the market for organic foods far outpaces the growth of the market for conventional foods.

One reason the public chooses to buy organic foods is the guarantee that food labeled "organic" contains no GE ingredients. Since GMOs do not require labels, consumers have no way of knowing which foods contain GE ingredients. This lack of labeling contributes to people's fears about GMOs. For years, consumers have been pushing for required labels, and food companies have been pushing back. As recently as 2015, Congress ruled that food companies are not required to label which of their products contain GMOs. In 2016, Vermont passed its own law requiring these labels, but President Barack Obama signed a different law the same year, so it never went into effect. However, the Safe and Accurate Food Labeling Law, which Obama signed in 2016, allows companies to put labels with QR codes on them that, when scanned with a smartphone, will link customers to the company's website for full information about the food.

The Safe and Accurate Food Labeling Law of 2016 gave food companies the option of putting a QR code similar to this one on their products. Scanning the code with a smartphone takes customers to the company's website to get more information about the GMOs in their food. This is voluntary, so not all GE foods have a QR code on them.

According to some, the lack of mandatory GMO labeling takes away consumers' right to make their own food choices. Others disagree, saying the 2016 law is the best policy. According to economics professors David R. Just and Harry M. Kaiser, mandatory labels could mislead, rather than inform, customers because a small label "provides just enough information to scare the uninformed, and is not based on any sound science. It doesn't answer questions regarding what was modified, why the modification was introduced, or what technologies were used prior to modification."[25] Therefore, they wrote, adding the QR code to packages would allow more space for these questions to be addressed and might make people feel better about GMOs. This debate will likely continue for years, but as of 2018, people can avoid GMOs by purchasing food with labels such as "GMO-free" or "certified organic." In most cases, though, organic food is far more expensive because, compared to GE crops, organic ones grow more slowly, produce less at one time (leading to more demand than some farms can meet), and require increased shipping costs in an effort to stop organic food from mixing with non-organic food. Some argue that if GMOs were banned completely, people who live in poverty would have an even more difficult time affording food.

A WAR AGAINST NATURE?

"There is a fundamental question here: Is progress really just about marching forward? We say no. We say it is time to stop assuming that discoveries only move us forward. The war against nature has to end. And we are going to stop it."

–Peter Melchett, former executive director of Greenpeace, UK

Quoted in Michael Specter, "The Pharmageddon Riddle," *New Yorker*, April 10, 2000, p. 58.

Leaping Hurdles

Though the FDA does not consider public opinion when it makes decisions about the safety of food, public opinion does impact research. Since the goal of biotech companies is to market their products, they must consider whether future profits justify the expense of the research. David Stark, a former vice president at Monsanto, estimated that bringing a new product from research and development to market takes an average of 10 years and $100 million. Many obstacles must be overcome before GMOs hit the shelves.

Aside from rulings by the FDA, biotech companies must overcome public resistance—sometimes called the "yuck factor." Those who oppose GMOs voice concerns about the environment and fears that once unleashed, GMO foods will overrun the ecosystem, eliminate biodiversity, and create super strains that choke out the competition. While farmers and companies both work to keep cross-contamination to a minimum, it is difficult to avoid completely, especially in canola and corn crops.

Impacting the Environment

The effects of GE plants and animals on ecosystems can be far-reaching. Not only are scientists beginning to understand that genes interact with each other in unpredictable ways, they also cannot predict how each ecosystem involved may be affected. For example, a study published in the *Proceedings of the National Academy of Sciences* by a team of environmental scientists in the Midwest found that water runoff from fields of GE Bt corn is washing into streams near other cornfields. The high levels of the toxin intended for the corn borer is also reducing the population of the caddis fly—an insect related to the corn borer. The reduction of caddis flies, in turn, deprives the fish and amphibians that feed on this insect. The birds that feed on the fish and amphibians also suffer. Not all scientists, however, accept this perspective.

In 1999, the public was outraged by a report in the highly respected scientific journal *Nature* that populations of

Although initial reports predicted terrible consequences for monarch butterflies that ate GE pollen, the actual impact to the population was minimal.

monarch butterflies were threatened by eating milkweed dusted with pollen from nearby Bt corn. The newspapers were filled with grim predictions about the destruction of the butterfly. Three years later, new studies found that the danger to the butterfly was actually minimal in real-world conditions outside the laboratory. However, that information was barely noted by news organizations and ignored by the public.

The monarch butterfly story illustrates how difficult it is to find the truth when each side has so much to gain from its perspective. The Pew Initiative on Food and

Biotechnology, a nonprofit, nonpartisan organization that concluded its work in 2007, published a report in 2002 titled "Three Years Later: Genetically Engineered Corn and the Monarch Butterfly." The report examined the background of the monarch butterfly story and how timing, special interests, science, politics, and industry combined to magnify a small story into an international sensation. The public was left with the impression that monarch butterflies are threatened, even though later studies—not as well publicized—did not replicate those findings. The report also highlighted how "eventually, with the leadership of the U.S. Department of Agriculture (USDA), scientists from government, industry, academia, and environmental groups worked together to develop a consensus set of experiments needed to answer the question of whether genetically engineered corn posed a risk."[26] It detailed the process of how the groups worked together so that each was satisfied that its concerns were addressed.

Hard to Resist

The passionate objection to GE foods may be showing signs of softening, and science and politics are only partly responsible. As food and energy prices rise, many countries can no longer afford the luxury of boycotting products that contain GMOs. When leaders must decide between philosophical objections and real hunger, their choices are limited.

Much of the world's corn crop lately has been diverted into producing biofuels, resulting in shortages of corn crops grown for food. Prices of rice and wheat have doubled in recent years, also resulting in global hardship. In some countries, farmers cannot afford to buy the rice they grow and sell to others.

Strict regulation of GE crops also creates hardships for livestock producers because of the rise in the cost of feed. Feed makers and livestock producers in Europe challenged strict rules in 2007 that forbid the import of grain from the United States if it contains even trace amounts of GE products. They asked that these tight restrictions be relaxed when small amounts of GE corn turned up in tests of corn gluten intended for animal feed.

A 2008 article in the *New York Times* stated that Japan and South Korea have resorted to buying GE corn for use in soft drinks, snacks, and other foods, since prices for conventional corn have tripled in recent decades. Steve Mercer, a spokesman for U.S. Wheat Associates, a federally supported cooperative that promotes U.S. wheat abroad, said, "I think it's pretty clear that price and supply concerns have people thinking a little bit differently today."[27] Though outspoken interests are at work on all sides of the GE issue, food shortages and rising prices have begun to chip away at longstanding resistance.

As these examples make clear, the most widespread challenge in genetic engineering, as in many new areas of science, is distinguishing the science from the politics. As the report from the Pew Initiative on Food and Biotechnology explained, it is difficult to sort out the tangled and complicated web of interests that control how, when, and if GE foods will gain acceptance by the general public.

Basic mistrust of new science and suspicions about the motivations of large biotechnology companies partly fuel the concerns of those who oppose GE foods. While those concerns may not be voiced outright, they underlie much of the criticism leveled at these products and the companies that promote them. Freeman J. Dyson pointed out that people might react differently to GE foods if the field were not driven by huge biotech companies overseen by government agencies.

He compared the current controversy over genetic engineering to mistrust of the early computer industry. In the 1970s, computers were mysterious and massive. They occupied entire buildings on university campuses; students would turn in punch cards with data in the morning and return in the afternoon to pick up their results. Dyson noted that in a similar way, large pharmaceutical companies and agribusinesses such as Monsanto drive and pay for most genetic engineering research. He reminded readers that the public began to accept computers as they became more familiar and user friendly. It was not that long ago that consumers eyed frozen foods and cake mixes with suspicion, too. Some people previously believed that advances in food preservation, such as refrigeration and canning, spelled doom for good cooking and fresh foods—which, of course, did not happen. While GMOs have currently been deemed safe, more research is needed in certain areas to prove this beyond a shadow of a doubt. However, the question may not be fully settled for years.

Fabric and Fuel

Although it is understandable that people want to know about the safety of their food, many have different concerns about GMOs that are not used for food. In those cases, the questions center less around the safety of the product and more on the alleged inability to contain GE organisms, the dangers they may pose to the environment, and fear of venturing into uncharted frontiers.

GE products can now be found on fashion runways and inside fuel tanks. As the technology advances, the applications widen. Sometimes questions are about aesthetic factors: Do people want to wear clothing made from genes that do not match those in nature? Sometimes the questions involve right and wrong: Is it ethical to use food to fuel factories and automobiles while people go hungry? These questions occupy the thoughts of scientists, politicians, and the public, who struggle to reconcile interests that may be in conflict. Health, profits, and the search for alternative energy are issues that continue to drive genetic engineering technology.

Spiders and Goats

Though the nearly transparent threads of a spider's web seem fragile, that delicacy is an illusion. Spider silk is actually one of the world's strongest natural materials. Its strength relative to its size is five times that of steel. In other words, if a strand of spider silk were as large and thick as a steel beam, it would be five times stronger.

The biotech industry hoped to use that strength for profit by engineering the gene responsible for spider silk into the DNA of goats. When the goats mature, they secrete the protein responsible for spider silk in their milk, creating a product called "milk silk." In this way, they produce much more silk than spiders. *Business Insider* explained,

> *And, it should be noted, the goats don't excrete a single fiber—that would be weird.*
>
> *The milk has to be separated and refined several times, then washed, freeze-dried, and turned into a powder. The powder can be spun into a fiber, or transformed into a coating or adhesive. (The milk isn't ever kept for human consumption.)*[28]

The silk this produces is only one-half to two-thirds as strong as natural spider silk, but this is still much stronger than other materials available today. The downside is that the process is expensive, making it unlikely that spider silk will ever become widely used. However, some organizations are willing to pay more for this higher quality product that has certain specialized uses.

Spider silk is five times as strong as steel, but it took genetic engineering to be able to produce it on a large enough scale for practical applications.

The goats were created by Nexia Biotechnologies, a company that went out of business in 2009. The goats were moved to Utah State University, where research continued. According to biologist Randy Lewis, the scientist who created the goats for Nexia, the animals have reproduced about eight or nine times and passed down their spider silk gene. In December 2015, Utah State University was given $1 million by the United States Army to use the silk to create body armor for soldiers.

Current lightweight body armor is made either with nylon or with a material called Kevlar. However, Kevlar is heavy, and nylon melts at high temperatures. In contrast, spider silk is lightweight and does not melt. According to Lewis, the silk has other possible uses as well. For example, the "human body doesn't reject the fiber like it does other materials. The silk could be used for sutures [stitches], skin grafts, or for complex jaw repair surgery."[29] It could also potentially be used to make ultra-strong bridges and cars.

Some people believe it is wrong to combine animal genes this way. Others, including Lewis, believe it is no different than the way people have been selectively breeding their animals for centuries, hoping to naturally get cows that produce more milk and dogs that have specific traits, such as fur of a certain color or better herding ability. Additionally, the goats are not harmed by the addition of the spider gene. The spiders themselves, however, do not fare so well. Two common spiders—a garden spider and a golden orb weaver native to the tropics—are used to make these goats. The spiders are flash frozen in liquid nitrogen and then ground into powder. The silk-producing genes are then extracted from the spider powder. These genes are engineered with a genetic switch that is programmed to turn on only when excreted in the milk of a female goat. Finally, the new genes are implanted in the eggs of female goats. Since the DNA of a goat consists of 70,000 genes, goats with spider genes are only one-seventy-thousandth spider. Once the goats give birth, the spider genes switch on, and the goat milk is fortified with spider silk protein.

Organic Fibers

Cotton is the crop with the greatest impact on the fashion industry. By the end of the first decade of the 21st century, half of the world's cotton crop consisted of Bt cotton—genetically engineered, like corn and soybeans, to resist pests. Though this might appear harmless enough, critics fear that controlling strains of harder-to-kill pests that have resulted from the use of these crops will require ever-stronger poisons. Aside from the danger to food crops growing nearby, these poisons might

Dragon Silk and Glowing Silk

Researchers have found out the hard way that it is difficult to harvest silk from actual spiders. In addition to the fact that they do not produce much at one time, spiders tend to eat each other when they are put in large groups. Silkworms, on the other hand, can live together peacefully. For this reason, researchers have implanted spider genes into silkworms to make their silk, which has been used for years to make silk clothing, stronger and more durable. When the silkworms reproduce, they pass on this trait. A company called Kraig Biocraft Laboratories hopes to use this silk, which they call Dragon Silk, to make body armor for soldiers as well as athletic clothing for regular citizens. According to Jon Rice, the chief operations officer (COO) at Kraig Biocraft Laboratories, Dragon Silk is about 1,000 times cheaper to manufacture than its competitors.

Another way people have genetically modified silkworms is by implanting them with genes from sea organisms, such as corals and jellyfish, that are fluorescent, which means they glow. The silkworms then spin fluorescent silk, which has been used to make glowing clothes such as wedding gowns, suits, and ties.

Scientists have found ways to engineer silkworms to make their silk have different properties.

endanger the health of workers in the field. Cotton is grown in many countries where workers are not protected by laws about their exposure to toxic chemicals. Ten million people in Central and West Africa depend on growing and processing cotton.

Cotton is one of many crops to be genetically engineered to resist pests.

The debate over organic versus conventional, or GE, cotton can be confusing. One of the things people consider is which one is better for the environment. Many say organic cotton is better because it uses less water and enriches the soil. However, others say these claims are misleading because conventional cotton has been engineered to produce more cotton per plant, meaning fewer plants need to be planted and watered. This also means they use less land space. In countries such as India, where water is scarce, this is an important concern.

Additionally, because organic cotton produces less, getting enough cotton from organic plants to meet the world's demand can be difficult. The clothing company prAna, for instance, is a strong supporter of organic cotton, but about 16 percent of the cotton it uses is conventional.

While many people are concerned about the environmental impact of different types of cotton, others believe the farmers are the most important concern. Organic cotton farmers use fewer pesticides, and the ones they do use are less harmful to them. Chemicals in non-organic pesticides, such as Monsanto's

Roundup, have been linked to cancer when farmers breathe them in. Companies such as prAna believe using organic cotton is important because it does not negatively affect the health of the people who grow it.

The website Quartz noted that there are other concerns people need to think about when trying to buy the most environmentally-friendly cotton clothing, such as what kinds of dyes were used: Non-organic dyes create a lot of waste water and use a lot of chemicals, and it is almost impossible for someone to figure out whether or not the dyes used in their clothing were organic. Quartz offered this advice to people who want to have as little negative impact as possible on both the environment and the people who are involved in producing clothing:

> *Buy better clothes. Buy less of them. Wear them more. There's no more certain way to reduce your impact than to reduce the amount of clothes you consume and to keep those clothes for a long time. Just try to wash them only as needed. Laundry uses a lot of water and energy, too.*[30]

Taking Precautions

Other clothing companies are also lining up to oppose genetic engineering, even if they do not have a direct connection to GE products. In 2002, Patagonia, a major outdoor clothing supplier, publicly criticized Bt corn and Roundup Ready soybeans. Yvon Chouinard, the founder and owner of Patagonia, issued a statement in which he said,

> *For humans to adopt every new technology just because we can is irresponsible, especially when a new technology puts us up against an edge that's hard to see, feel, or even define. New technologies, like genetically engineered food, should be considered unsafe until proven otherwise, not the other way around.*[31]

In the same essay, Chouinard explained the "Precautionary Principle," which states that if an activity might threaten safety, precautionary measures should be taken even if not all cause and effect relationships are fully understood. In other words, the burden of proof lies with the individual or the company that is taking the action, not with the public that might suffer unknown consequences.

Genetics at the Gas Pump

In addition to playing a role in creating clothing, genetic engineering is helping companies find alternatives to fuels made from materials such as coal and oil. These fossil fuels release carbon dioxide, which contributes to global warming and climate change. Additionally, they must be taken from the ground through drilling processes that damage the land and kill plants and animals. For these reasons, biotechnology companies are genetically engineering corn and other crops to make it easier to produce ethanol and other biofuels, which release fewer greenhouse gases and eliminate the need for drilling. Ethanol is a form of alcohol similar to that found in alcoholic beverages. In the United States, ethanol is generally made from corn, though other countries, such as Brazil, use sugarcane. Soybeans and prairie grass are also used.

The process of converting corn into ethanol involves the addition of amylase, an enzyme, at the ethanol factory. Amylase breaks cornstarch down into sugar, which is then fermented and processed into ethanol. A biotechnology company called Syngenta has genetically engineered a type of corn called Enogen that eliminates the need to add amylase by adding it directly to the corn.

Even gas companies have jumped on board. ExxonMobil, the largest oil and gas corporation in the world, is currently working with biotech company Synthetic Genomics to create biofuel using GE algae. By 2025, they may be able to produce 10,000 barrels of this fuel per day.

People use a combination of gasoline and ethanol in their cars today, but powering vehicles with nothing but corn, bacteria, or algae still raises some issues. An article in the *New York Times* explained one of the issues that faces ethanol supporters: "Even if the nation's entire corn crop were converted to ethanol production, it would replace only about 15 percent of petroleum use, according to an Energy Department report."[32] This is because biofuel is less efficient than gasoline, so more is needed to

Algae—organisms related to seaweed—are versatile and used in many genetic engineering projects.

make a vehicle travel the same distance.

Politicians from rural farming states have joined the push to grow corn for fuel, since the effort raises local incomes and benefits their states' economies. The National Corn Growers Association and the Renewable Fuels Association express belief that record harvests ensure enough corn for every purpose. However, as with any new technology, it is impossible to plan for every possibility. Each scientific advance has unforeseen results, some of which are positive and others negative.

Another fear is the same that critics voice about any technology that involves GE food crops—the possibility that the corn that has been modified for ethanol production will accidentally contaminate food crops. In fact, this happened in 2017, when some Enogen corn pollen was blown into a field of non-GE corn. Although the corn is not dangerous for people to eat, the extra amylase in it means its starches break down when it is processed, so corn chips fall apart and cereals get soggy quickly.

One biotechnology company is working on a solution to the problem of cross-contamination. Sustainable Oils is a joint venture by two companies: Green Earth Fuels, which manufactures biodiesel fuel, and Targeted Growth, which develops GE plants. Targeted Growth has genetically engineered the camelina plant to produce seeds with 20 percent more oil than seeds from conventional plants. That oil is then processed into biodiesel fuel. Furthermore, although the camelina plant is a distant relative of canola, a plant used to make cooking oil and other products, humans do not eat camelina itself, which means that they will not unintentionally ingest it. Additionally, since it thrives on parched, dry land, the crop benefits farmers who live in regions that depend on artificial irrigation to water their crops.

Economics and Biofuels

Because they are made from food, biofuel production has a direct impact as well as several indirect effects on food supply. According to Chris Smith, the cofounder of a science podcast called *The Naked Scientists*,

> *In the case of US corn, for example, using the crop to produce fuel means that the price of corn to eat goes up. So what, you might ask, I can live without sweetcorn? But corn is widely used as an animal feed, so the price of beef goes up. Livestock farmers then look for alternative feed and the price of other grains–wheat, barley or rice–goes up. So using corn for fuel makes the meat and the bread in your burger more expensive!*[1]

This increase in prices may not mean much to most people in developed countries, "but in the developing world higher food prices can be catastrophic. Food riots and political instability can result."[2] Riots have already occurred in Mexico over the rising cost of corn tortillas. In early 2018, the National Tortilla Council warned that it could raise prices up to 20 percent, which triggered more outrage and threats of protests on social media. Mexican economic officials called the price raises unjustified and agreed to work with the Tortilla Council to keep costs down.

1. Chris Smith, "The Trouble with Biofuels," *The Naked Scientists*, April 24, 2013. www.thenakedscientists.com/articles/science-news/trouble-biofuels.
2. Smith, "The Trouble with Biofuels."

Biodiesel fuel has become popular in recent years as consumers have become increasingly aware of the dangers posed by carbon dioxide emissions. Biodiesel fuel produces almost 80 percent less carbon dioxide than regular diesel. Moreover, most vehicles that use regular diesel fuel can use biodiesel with minimal modifications. Ethanol, on the other hand, is far less efficient. Ethanol must be mixed into gasoline in various concentrations, and most cars cannot run on ethanol alone.

A Look at the Future

In July 2008, Chris Ladd, a writer for *Popular Mechanics* magazine, interviewed geneticist J. Craig Venter, the cofounder of Synthetic Genomics. Venter helped in the first efforts to map the human genome and began a project to produce biofuels by genetically engineering algae. Ladd asked Venter what he thought the near future of genetics would be, to which Venter replied:

> *I see it as parallel to the electronics industry in the 1940s and '50s, a stage when all of the things that enabled computers came out of just a few handfuls of components—resistors, transistors, capacitors—and people were pretty much limited only by their imagination. My team has discovered more than 20 million new genes, so we're in a biological universe. There are no fundamental limits. I think we're going to see the next 25 years as some of the most innovative in the history of science.*[1]

Another nonfood-to-fuel solution may be the cultivation of certain grasses, such as switch grass, which, like camelina, is low maintenance. Genetic engineers are working on developing strains of plants with less lignin, the substance that makes plants stiff and also interferes with the process of converting the plant into ethanol. This prospect, however, alarms environmentalists who worry about cross-pollination.

The applications of genetic engineering have just begun to show up in food, drugs, fashion, and fuel. Someday humankind may look back at the beginning of the 21st century and wonder why people worried about some of the things that now concern them and missed some of the problems that no one imagined. Dyson outlined the five questions that he believes must be addressed as biotechnology moves from the outward fringes of science into the everyday lives of ordinary people: "First, can it be stopped? Second, ought it to be stopped? Third, if stopping it is either impossible or undesirable, what are the appropriate limits that our society must impose on it? Fourth, how should the limits be decided? Fifth, how should the limits be enforced, nationally and internationally?"[33] Dyson did not answer these questions. Instead, he wrote, he leaves the solutions to future generations. Dyson believes that everyone shares in the responsibility to understand genetic engineering and to voice opinions about how this powerful technology is applied.

CHAPTER 4

Designing Babies

Even among people who support using GMOs for food, fuel, and fabric, there is fear and distaste surrounding the idea of genetically engineering humans. Some believe there is nothing wrong with it at all, while others feel it is wrong no matter what. Some people fall in the middle, favoring it in certain circumstances; for example, they may be fine with editing genes to eliminate the possibility that someone will develop an inherited disease but strongly opposed to adding or changing genes identified with certain traits. Those traits could be cosmetic, such as eye and hair color, or involve abilities and talents, such as musicianship, intelligence, or athleticism. How much control should humans have over their children's genetics? Is it fair? What will the future be like if some people are genetically enhanced and others are not? Dyson's questions about the future of genetic engineering become most urgent when considering the lives of human beings, particularly since these decisions affect the lives of children.

A New Fashion Trend

In 2018, the first baby conceived in a test tube celebrated her 40th birthday. In 1978, the birth of Louise Brown made headline news around the world. Childless couples desperate to have a family had newfound hope. However, the rapidly growing science of genetics may soon provide parents with options that no one even imagined 40 years ago.

One day, people may be able to choose how tall their child could become, what color eyes and hair that child will have, or even how intelligent and talented they may be. A 1999 article in *TIME* magazine about "designer babies" compared the future possibilities to a shopping trip: "Before the new millennium is many years old, parents may be going to fertility clinics and picking from a list of options the way car buyers order air conditioning and chrome-alloy wheels."[34] This tendency to compare the selection of a baby's genetic traits to a shopping trip, and even the term "designer babies" itself, hints at genetic engineering's greatest obstacle: Many people do not believe it is right to pick and choose the details of a child as if they were a fashion accessory.

In the future, parents may be able to choose their baby's traits, making them genetically predisposed to being taller, smarter, or better at sports.

While the topic of designer babies often arises when genetic engineering is discussed, the reality is that adding to or changing the genes in an embryo's sperm or egg cells—known as germ cells—are new and unexplored advances. Germline manipulation is the scientific term given to changing an embryo's DNA so that the embryo will inherit traits that might then be passed along to future children. This is the genetic engineering people most associate with designer babies. Somatic cells—the ones that make up most of the body, such as blood and brain cells—cannot be passed down, so any changes a person makes to their own body will not change future generations.

Critics, however, voice serious concerns. Society places great importance on unconditional parental love. In 2004, Michael Sandel, a Harvard University philosopher, published an essay on the ethics of genetic engineering titled "The Case Against Perfection." In that essay, Sandel weighed parents' rights to make genetic choices against the dangers of making such choices. Sandel warned that society suffers when children are viewed as products, with some children—and some traits—considered more valuable than others.

Other writers also refer to Sandel's essay to help clarify their ideas about genetic engineering. Mike Polyakov, formerly a graduate student at the University of California, Berkeley, for example, disagreed with Sandel's belief that parents should be given a choice about some aspects of their children's genetic makeup—such as predisposition to certain diseases—but not others. Polyakov wrote, "One can't condemn parents for choosing a smarter or taller than average child on these grounds, for they can reasonably say they are doing it entirely for the child's good."[35]

For the sake of argument, he assumed that everyone has equal access to these genetic benefits. He also pointed out that what people consider "normal" varies. A condition considered normal in the jungles of the tropics, for example, may be abnormal in the Arctic. Abstract concepts such as beauty and intelligence also vary widely from one culture to another.

UNLIMITED POSSIBILITIES

"Anything that I can imagine in terms of changing genes in a baby, I could do. I could give a baby the hearing ability of a dog or the eyesight of a hawk."

–Lee Silver, geneticist

The Ethics of Breeding Better Babies

Another issue surrounding genetic engineering is eugenics, a social movement that was promoted mostly in the 19th and 20th centuries. Eugenics involves what some people see as an improvement of the human race by allowing only those with desired genetic traits to bear children. It is considered an extreme example of how genetic engineering might be abused.

Eugenics advocates have often promoted the sterilization of people who have traits considered undesirable. Moreover, they favor the deliberate selection of mothers and fathers to breed superior children. Today, most people associate eugenics with Nazi policies during World War II. At that time, the Nazis killed many people who did not have traits they considered ideal, such as birth defects, disabilities, and mental illness. However, according to PBS, Hitler was inspired by an American eugenics program that was started in California in 1909. This program performed "sterilization procedures … on men and women, often without their full knowledge and consent."[36] Other states had similar programs. California's forced sterilizations included "immigrants, people of color, poor people, unmarried mothers, the disabled, [and] the mentally ill"[37] and continued almost until the 1980s. Other eugenics programs were active even longer, such as the one targeting Native Americans. According to the PBS article "Unwanted Sterilization and Eugenics Programs in the United States," 25 to 50 percent of all Native American women were forcibly sterilized between 1970 and 1976 alone. Some eugenics programs may still be secretly operating today.

Eugenics programs were designed to stop people with "undesirable" traits—some of which were not genetic, such as poverty—from having children. They were especially active in U.S. prisons because people in the early 20th century falsely believed lawbreaking was hereditary, but eugenics programs were also practiced in regular hospitals around the country. Women would go in for a routine medical procedure such as an appendectomy (an operation to remove the appendix) and receive tubal ligations at the same time without their consent. A tubal ligation, often referred to as a woman "getting her tubes tied," is an operation in which the fallopian tubes, which send an egg from the ovaries to the uterus, are cut, tied, or otherwise blocked. If the egg cannot get to the uterus, no pregnancy can take place. Some women undergo this procedure voluntarily as a permanent form of birth control, but in eugenics programs, the surgery was not voluntary. A tubal ligation can sometimes be reversed, although this requires further surgery and would likely have been too expensive for many of the targeted women. Other methods of eugenics, such as removing the entire uterus, were not reversible.

A PERSONAL ACCOUNT

"I did not find out that they had sterilized me until I was 19 years old. I asked the state of North Carolina why they did this to me and they said that because I was feebleminded [outdated term for mentally challenged], that I would not be able to take care of myself ... They were saying that feeblemindedness is hereditary, so they sterilized me so I would not produce 'my kind.' Mind you, I am not illiterate nor am I feebleminded. I never went into high school, but yet I still acquired a college degree."

–Elaine Riddick, victim of North Carolina's eugenics program

The threat of eugenics makes many people fear what might happen if genetic engineering of future children becomes widespread. There is already controversy around this; in some cases, pregnant women who are told their child may have some sort of mental or physical defect choose to have an abortion. Some people believe this is acceptable, while others believe it is similar to the Nazis' "racial hygiene" policies. Polyakov offered a compromise. He suggested that individual parents should be free to make choices for their own children, while denying those same choices to society as a whole—for example, he does not support a mandatory government genetic engineering program. He reminded readers that genes simply play one role in determining how a child will develop. They do not absolutely determine talents, intelligence, or athletic ability. Parents, siblings, friends, education, and the place where someone is raised are among the many factors that contribute to a child's development.

Fixing Genes

Regardless of how people feel about the issue, the possibility of genetically engineering "superbabies" remains a figment of popular imagination. Most of the current work in genetic engineering involves manipulating DNA to fix disorders that result when a person's genes malfunction. Parkinson's disease, Alzheimer's disease, certain cancers such as leukemia, and diabetes are among the diseases that result when systems regulated by genes stop functioning correctly. Most genetic conditions result from the combined effects of several genetic malfunctions.

Several techniques are used for repairing malfunctioning genes. One technique, known as somatic cell manipulation—or more commonly, gene therapy—involves introducing genes into the body to correct faulty messages that certain genes communicate to cells. Genes are actually coded instructions for building proteins, which then perform all the functions within the body. Sometimes, a mutation in a gene or several genes causes those instructions go haywire. Tumor growth, for example, is the result of a failure of cells to stop growing. Some genetic mutations are not harmful; for example, red hair is the result of a mutation in a gene called MC1R. Others, however, cause diseases.

Gene therapy can take different forms. Sometimes abnormal genes are swapped for healthy ones in the hope the healthy genes will take over and restore normal function. Other times, doctors might actually repair the abnormal gene. It is even possible to chemically adjust the degree to which a gene is turned on or off. An overactive gene can be stabilized, or a slow gene can be chemically induced to function more efficiently.

Gene therapy that involves replacement genes presents additional challenges, however. These replacement genes, which act as tiny repair kits, are carried into the cells by a molecule—generally a virus—that also has been genetically altered to accomplish its task. Some fear that the altered viruses may recover their original ability to cause disease and harm immune systems that are already weakened. People sometimes develop immune responses to the viruses that carry the new genes. Furthermore, since cells constantly replace themselves, replacement genes must be replenished frequently, which makes gene therapy short-lived and therefore somewhat impractical. As of 2018, it is being tested only for diseases with no other cure because of the risks involved in the treatment.

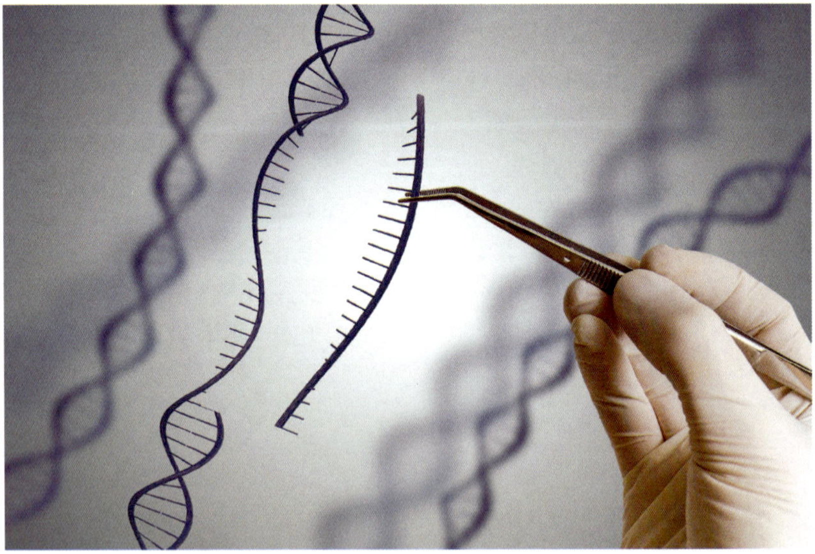

Gene therapy replaces or repairs malfunctioning genes, but the results are sometimes difficult to maintain.

Because gene therapy is largely experimental, it is not widely used to treat genetic disorders. For a long time, the FDA would not approve gene therapy, but in August 2017, a therapy to treat children and young adults with leukemia was finally given the green light. In December of that year, a treatment for an inherited eye disease that causes blindness was also approved. In 2018, a 13-year-old boy named Jack Hogan was the first to receive this gene therapy for blindness and became part of medical history in the United States.

Each time scientists identify a genetic variation linked to a disease or disorder, they move closer to learning how to engineer the genes to fix the problems. Successes do not, however, necessarily result in widespread approval. The public and scientists struggle with questions such as: What is normal? What is a disorder? Who decides? Is it ethical to devote so many resources to finding cures for diseases that afflict some people when millions of others suffer from problems such as malnutrition that the same resources might prevent? The answers to these questions are anything but clear.

Doping Genes

In 1999, the World Anti-Doping Agency (WADA) was formed with the goal of keeping performance enhancing drugs out of sports. Back then, gene doping, the term given to gene therapy intended to enhance athletic performance, was an unknown problem, but it has since become a potential issue. Unlike performance-enhancing drugs, gene doping involves transferring genes rather than drugs into an athlete's cells or tissues. Because the new genes match the athlete's own, they are undetectable by standard tests. In fact, the only way right now to detect gene doping is to study the secondary effects on an athlete's body. WADA works hard to stay informed about each new technique devised to cheat the system in competitive sports. Gene doping and gene editing have been banned in professional sports, and in 2016, athletes in the Olympics were tested for gene doping for the first time. However, most researchers believe the main concern for WADA and other sports organizations should remain drugs such as steroids, not gene enhancement, which is unlikely to become widespread any time soon.

Living Forever

While engineering genetic improvements in babies and adults fascinates the public and scientists alike, the possibility of immortality, or living forever, is a subject that has captured the human imagination since ancient times. The gods and goddesses of ancient Greece and Rome, for example, appeared human but possessed superhuman powers, including immortality. Cynthia Kenyon, the vice president of Calico, a lifespan research company, is studying how genetic engineering may actually turn that dream into reality. In 1993, Kenyon and her team found that changing a specific gene in roundworms doubled the life span of these parasites, which normally live only about three weeks. Their brief life span makes roundworms ideal subjects for aging studies, since changing their genes shows quick results.

The discovery that genetic changes affect life span is monumental because most scientists had assumed that aging is inevitable—the result of cells no longer renewing themselves. In other words, scientists used to assume that the body simply wears out as time passes. However, further studies on other lab animals, including mice, confirmed Kenyon's findings that genes can be changed to extend life span. Furthermore, the same gene that slows aging also seems to protect against cancer and tumor growth. Some scholars predict that human life expectancy could increase to 5,000 years in rich countries within the next 100 years, though others believe this is an idea that will remain in science fiction.

If this technology could successfully be created, many people would likely want to have it. However, the ability to control aging may have unintended and potentially devastating consequences to the global economy, health care systems, governments, and the environment. Even today, some experts say the planet is overpopulated, so Earth would have a hard time supporting a population that keeps increasing. Nicholas D. Kristof addressed the issue in a *New York Times* editorial when he wrote, "We are now blundering toward genetic manipulations, a technology that we should embrace—but prudently. It will reshape humanity far more than fire, electricity, space exploration or any other branch of science we have encountered."[37]

Creating Chimeras

In Greek mythology, the Chimera is a creature made up of parts of three different animals. In genetics, a chimera is an organism with at least two different sets of DNA. It is different than a hybrid, which is an organism "containing genetically identical populations of cells originating from a cross of two different species."[38]

The Chimera is a monster from ancient Greek mythology. It is often drawn as a lion with a goat's head coming out of its back and a snake for a tail.

Scientists around the world conduct research on chimeras so they can study human diseases without experimenting on humans. At Stanford University in California, for example, pathologist (a doctor who studies diseases) Irving Weissman transplanted human brain cells into the brains of mice to study brain cancer. In 2005, Weissman's chimeric mouse brains were 99 percent mouse and 1 percent human.

Genetic Engineering in Fiction

Books have been created around the topic of genetic engineering for decades. Aside from the ChickieNobs in *Oryx and Crake*, Margaret Atwood's characters create animals such as pigoons (pigs growing human organs for transplant patients) and rakunks (a cross of a raccoon and a skunk). Eventually, one of the characters creates an entirely new race of humanlike people. In the young adult novel *Dr. Franklin's Island* by Ann Halam—based on H. G. Wells's *The Island of Doctor Moreau*, in which a scientist combines human and animal DNA—Dr. Franklin kidnaps the survivors of a plane crash and turns them into transgenic animals. Further back, Aldous Huxley's 1931 classic novel *Brave New World* describes a society where all the citizens are genetically engineered and developed in artificial wombs.

Movies have taken on this topic as well. Michael Crichton's 1990 book *Jurassic Park* has been expanded into the *Jurassic Park* and *Jurassic World* film series. The 2018 installment, *Jurassic World: Fallen Kingdom*, especially addresses genetic engineering with the creation of new dinosaur species and the dangers these species could present to the general public. In addition to film versions of books such as *Jurassic Park*, original films have been made about this topic. For example, 2009's *Splice* involves two scientists using human cells to create a hybrid creature in an attempt to become famous. Other movies include New Zealand's *Black Sheep*, in which genetically modified meat-eating sheep start attacking people, and *The Boys from Brazil*, about a group of Nazi scientists who try to clone Adolf Hitler nearly 100 times. According to a review of *The Boys from Brazil*, this movie is especially interesting because "it shows it's not only genetics [that] defines a person, but also the environment they grow in."[1]

1. Clara Rodriguez Fernandez, "The 10 Best Biotechnology Movies to Watch This Summer," Labiotech, July 25, 2017. labiotech.eu/best-biotechnology-movies.

However, he has proposed creating mice with brains that are composed entirely of human brain cells.

In other research, efforts to develop a ready source of organs and blood for human use have resulted in pigs with human blood in their veins, and researchers hope to go even further in the coming years. For decades, people have suggested using pigs to grow organs for transplant patients. Pigs and humans are so similar genetically that pig tissue has been used for years to replace human heart valves. However, they are different enough that the whole heart cannot yet be replaced. In medical trials, pig organs that were placed into monkeys have been rejected, causing the monkeys to die. The body rejects a new organ if the organ's genes are not similar enough to the host body's. The longest a monkey has lived with a pig organ inside of it—in this case, a kidney—was 400 days. However, researchers are encouraged by this because the previous record was only 150 days.

There have also been concerns over pig viruses being transferred to the transplant patient. However, scientists have used genetic engineering to solve this problem as well. A gene-editing tool called CRISPR allows researchers to remove certain things from genes while leaving the rest of the gene intact. CRISPR allows scientists to "eliminate immune-provoking sugars from the surface of pig cells, introduce human genes that regulate blood coagulation to prevent dangerous clots, and snip out viral sequences that some fear could infect a human host."[39] Scientists also hope to one day perfect CRISPR enough to eliminate genetic diseases in humans, but that day is a long way off. The pig-to-human transplant technology is also still far from being perfected, but as technology continues to advance, researchers remain hopeful.

What Percent Is Human?

Many people think it is a good thing researchers are trying to figure out ways to combine human and animal DNA to solve medical problems. However, these same people might feel differently if the changes to the animals were visible. Lee

Silver, a Princeton University geneticist, believes that people will accept these creatures as long as they do not appear human in any way. He pointed out that while people might accept a pig with a human kidney, they would most likely react in horror to a pig with a human arm.

External features are only one part of the debate, however. Many people fear that scientists will create an organism that does not have human features but can think and feel like a human being. If an animal such as a mouse had a completely human brain, such as that proposed by Weissman, would this mouse then deserve rights that are solely granted to humans? If the decision that an alert organism with the intelligence and feelings of a human being is not human because it does not look human, then how does society decide who—or what—is human? Silver believes that the boundaries are fuzzy. He compared the human-animal boundary to the color spectrum, in which one color undergoes very gradual changes until, finally, it is identified as another color.

These decisions become increasingly complex if the organism has human features. Though the scientific community shares the public fear of creating chimeras capable of human thoughts and feelings, the fact remains that this could happen someday. As Silver explained, "Modern biotechnology has moved such ambiguous beings from the realm of mythology to the realm of possibility. And this fact elicits the greatest challenge to western thought, which is that the existence of a strict line separating human beings from non-human beings may simply be a figment of our imagination."[40]

The boundaries between what is considered right and wrong have also been fiercely debated. However, politicians and the scientific community alike are not content to leave these questions open to individual interpretation. In 2006, President George W. Bush asked Congress to pass legislation banning the creation of human-animal hybrids since, he believed, they devalued human life. Many share Bush's concern. However, no U.S. federal laws currently address these issues.

The National Academy of Sciences assembled a committee in 2005 to provide voluntary ethical guidelines for chimera research. The committee, cochaired by biochemist Richard O. Hynes of the Massachusetts Institute of Technology (MIT), addressed two primary issues: engineering human cells into the germlines of animals to create inheritable traits and engineering brains that are partially human.

The first issue involves the production of human eggs and sperm by a nonhuman animal. If a male and female chimera with human eggs and sperm were to breed, they could theoretically produce a living human embryo. To prevent this possibility, the committee ruled that breeding chimeras is unethical. The rulings, however, have no regulatory power—they are merely recommendations.

The second issue targets research in which human cells are transferred into the brains of other animals, particularly primates, such as apes and chimpanzees. The committee also criticized these experiments. Richard Doerflinger, formerly of the U.S. Conference of Catholic Bishops, expressed the concern of many when he said, "If something were half human and half animal, what would our moral responsibilities be? It might be immoral to kill such a creature [by experimenting on it]. It's wrong to create creatures whose moral stature we are perplexed [confused] about."[41]

Doerflinger is not alone in his discomfort. Biotechnology activist Jeremy Rifkin objects on other grounds. He believes that animals have the right to exist without human tampering. He favors the use of computer models rather than live animals to study disease systems. Animal rights activists agree with Rifkin.

Other countries have been more proactive. Canada, for example, passed the Assisted Human Reproduction Act, which bans several things:

- creating a human clone
- transferring a nonhuman cell into a human embryo—or vice versa—if the embryo is expected to live
- paying for human eggs
- receiving payment for arranging surrogacy (when a woman has a baby for a woman who cannot have one)

- implanting a genetically altered embryo into a woman's uterus
- using genetic engineering to try to control a future child's sex
- genetically altering a human embryo in a way that allows the modification to be passed on to future generations

Creating chimeras from apes such as chimpanzees, one of humans' closest animal relatives, has been called highly unethical.

The law was created in 2004, but the first prosecution did not occur until 2014, when a woman named Leia Picard was convicted for paying for human eggs and receiving payment for arranging surrogacies. As of 2018, no one has been convicted for creating a chimera. Weissman, who created the mice with partially human brains, opposes a similar ban in the United States because he believes it would stand in the way of medical research that could save human lives.

However, like every other aspect of genetic engineering, the technology involved in the genetic engineering of humans moves quickly. Furthermore, it is not always clear from the outset just how a new science or technology will play out once it is released from the strictly controlled environment of the laboratory. Like the DNA that genetic engineering seeks to improve, genetic engineering itself interacts with the social, political, and cultural environment in ways that can be unexpected. Complex new ethical dilemmas arise even as society struggles to resolve the old ones.

CHAPTER 5

Extreme Genetic Engineering

Like a jigsaw puzzle, finding and fitting the first pieces of genetic engineering were the most difficult. Once the basic outline is complete, however, the remaining pieces tend to fall into place. As time marches on and technology continues to advance, humanity finds itself in a new but not unfamiliar position at the edge of a major scientific revolution. Genetic engineering technology has become and will continue to become cheaper and better, opening possibilities to both scientists and the public alike.

Minty Bacteria

Rapid advances in the understanding of how genes work as well as progress in the fields of engineering and technology have resulted in two fields of study that now capture a great deal of attention: bioengineering and biotechnology. Engineers focus on how things work and how to make them work efficiently; bioengineers apply those concepts to living things. Biotechnology involves the use of technology to make or change products for specific purposes. Biotechnologists, for example, develop new ways to treat waste, manufacture biological weapons, and make food. In 2017, for example, scientists created a new way of processing the Ruby cocoa bean. This processing technique created Ruby chocolate, which is naturally pink with a berry-like flavor.

Biotechnology is closely related to genetic engineering, since bioengineers apply the growing understanding of how genes work to solve some of the planet's most perplexing problems. Bioengineers are responsible for creating pest-resistant varieties of corn and soy and transferring genes between species to develop organs for transplant or ready sources of insulin.

Building with BioBricks

The International Genetically Engineered Machine Competition (iGEM) sponsored by MIT is a worldwide event that addresses the question, "Can simple biological systems be built from standard, interchangeable parts and operated in living cells?" In order to answer this question, teams of participants–mostly college undergraduates–are supplied with standardized biological parts known as BioBricks, which can be popped into cells like tiny LEGOs. Their challenge: design and build genetic machines that perform useful functions. Students from universities around the world participate in the competition. The interchangeable biological parts consist mostly of bacterial genes, such as *E. coli*, although some use plant, mammal, and stem cells. An article in the *Boston Globe* explained, "The goal of the competition is to get young scientists to help spawn new industries in synthetic biology and make sure scientists worldwide are involved."[1] The winner of the grand prize takes home a large aluminum BioBrick trophy.

In one example, graduate students in a bioengineering lab at MIT were tired of having to smell *Escherichia coli*, often shortened to *E. coli*, a bacterium they were using in their experiments. *E. coli* emits the distinctive odor associated with feces. Their solution: They created minty fresh, wintergreen *E. coli* by combining the wintergreen gene from petunia plants with *E. coli* bacteria. As a result of this genetic engineering feat, the *E. coli* in their lab then smelled like mint. They engineered the bacteria further by inserting a biological trigger that changed the scent from mint to banana when the *E. coli* stopped growing. They could then engage in leisure activities within smelling distance of the lab while they waited for the smell to change from mint to banana. This bioengineering feat might someday improve the odors from landfills, sewage treatment plants, and even armpits.

Changing E. coli *to smell like a banana and mint smoothie allowed graduate students to study the bacterium without suffering from its smell.*

Bioengineers regularly cross boundaries between life forms in their quest for new solutions to difficult problems. However, until very recently, their efforts were limited to genes that were already available. As the 21st century marches on, bioengineers are now learning to build genes from scratch to accomplish specific tasks. The ability to build new genes has given birth to a new field: synthetic biology. Synthetic biology, sometimes called "extreme genetic engineering," raises questions about the very meaning of life itself. Is life just a machine made of parts that can be assembled and reassembled in various combinations? Synthetic biologists think so.

A Worldwide Revolution

"If we have another huge industrial revolution, it's really important that it happens worldwide … We have to have people all over the world making materials and energy to meet their own needs."

–Randy Rettberg, president and founder of IGEM

Genetics Customized to Order

The root of synthetic biology is the fact that the genes in all living things are composed of sequences of only four chemical bases: adenine (A), guanine (G), cytosine (C), and thymine (T). The genes of plants, animals, and bacteria vary only by the arrangements of these four chemical bases in the DNA. More than 50 years ago, scientists created the first tiny snippet of artificial DNA by artificially linking these four crucial chemicals.

However, not until 2008 did scientists report they had created the entire genome of a bacterium, which was 10 times longer than any that had been previously synthesized. The *New York Times* called this a watershed event, meaning it would change the course of science. Once scientists become skilled at synthesizing entire genomes, that knowledge will enable them to construct entirely new forms of life with specific functions. Andrew Pollack, *New York Times* science reporter, explained, "Synthetic biologists envision being able to design an organism on a computer, press the 'print' button to have the necessary DNA made and then put that DNA into a cell to produce a custom-made creature."[42]

Many scientists hope that solutions to global problems that have so far defied science might lie in synthetic biology. The biotech company Synthetic Genomics built the first completely artificial bacterium and is known to be working to develop organisms that produce ethanol, hydrogen, and other alternative fuels with the help of ExxonMobil. These organisms would provide an alternative source of fuel as well as enrich the company that wins the first patent on the technology. Some estimate the market for this future fuel could exceed $1 trillion.

The development of alternative fuel is only one application that propels research in "synbio," as people in the field often call it. Jay D. Keasling at the University of California, Berkeley, has developed synthetic DNA that allows genes from the wormwood tree and yeast to work together within *E. coli* bacteria to produce a steady source of artemisinin, a drug used to treat malaria—a leading cause of death and illness worldwide. In addition, scientists at MIT have developed artificial gene circuits that can be inserted in the body's cells to guard against cancer. Like miniature early warning systems, these circuits trigger the body's immune system to attack cells as soon as cancer is detected.

GROWING A HOME

"'Grow a house' is on the to-do list of the M.I.T. Synthetic Biology Working Group, presumably meaning that an acorn might be reprogrammed to generate walls, oak floors, and a roof instead of the usual trunk and branches."

–Nicholas Wade, science writer and editor

Is It Safe?

Though many scientists and entrepreneurs foresee great promise in synthetic biology, the field is still quite new. For the time being, most synthetic biology remains only partly synthetic, relying on natural organisms to do the bulk of

Right now, synthetic biology generally requires both natural and artificial components to reach the desired outcome.

the genetic work. It has been described as "a cut-and-paste operation, like writing a phrase by snipping the necessary words out of magazines and gluing them together in the proper order."[13]

Not everyone expresses unconditional enthusiasm about synthetic biology, however. Many voice concerns that DNA synthesis might be used in sinister ways. Some fear the possibility of deadly organisms contaminating the environment in purposeful acts of bioterrorism.

In 2002, scientists at Stony Brook University in New York State announced that they had synthesized the polio virus using information that is readily available. In fact, they ordered the necessary base pairs over the internet. In England, a reporter for the *Guardian*, a British newspaper, placed an order for a synthetic fragment of smallpox DNA and had it delivered to his home. Smallpox is a highly contagious virus that disappeared almost entirely more than 30 years ago. Scientists have also recreated the virus responsible for the deadly flu epidemic that killed between 20 million and 50 million people worldwide in 1918 and 1919.

In 2005, the U.S. Department of Health and Human Services published the full genome of that particular flu virus on the internet as part of an article that detailed how the virus had evolved from simple bird flu to a deadly human epidemic. Two science writers, Ray Kurzweil and Bill Joy expressed outrage that such potentially dangerous information was so easily accessible. In an editorial titled "Recipe for Destruction," they wrote,

> *The genome is essentially the design of a weapon of mass destruction. No responsible scientist would advocate publishing precise designs for an atomic bomb, and … revealing the sequence for the flu virus is even more dangerous …*
>
> *We should … treat the genetic sequences of pathological biological viruses with no less care than designs for nuclear weapons.*[44]

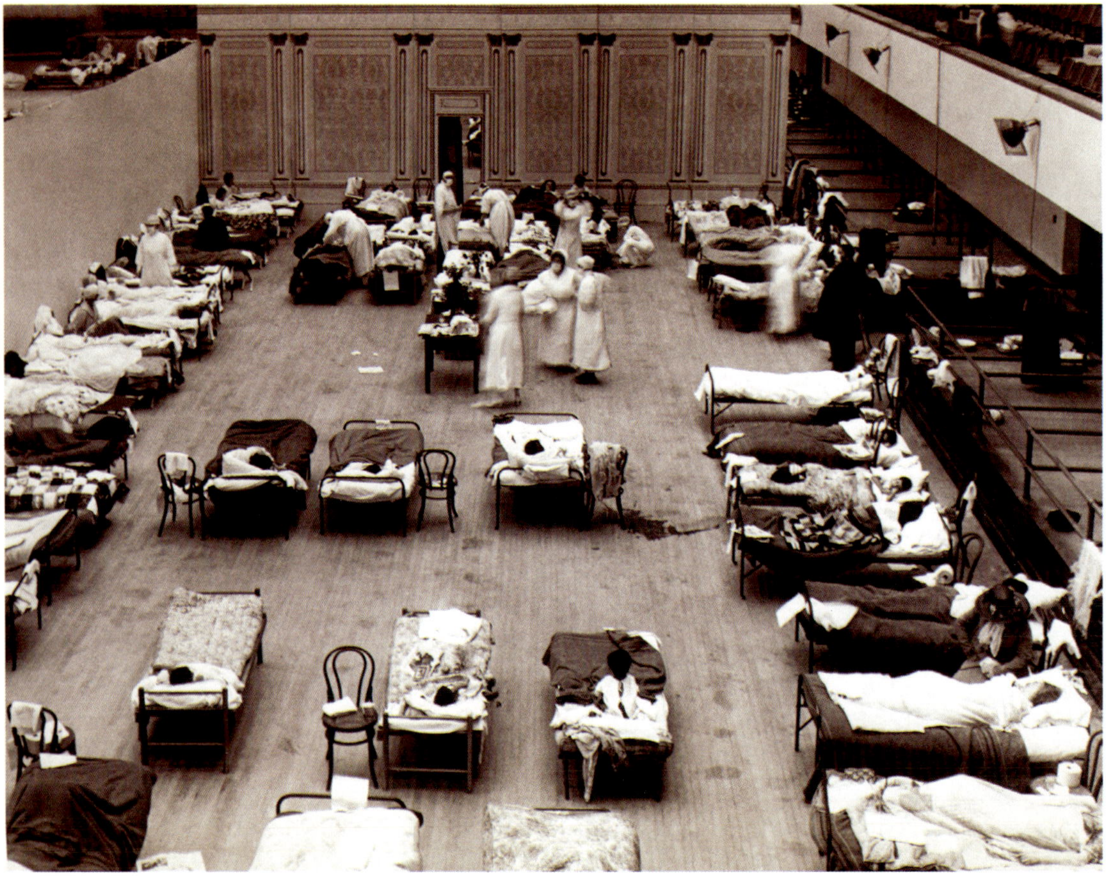

The 1918 strain of the flu virus caused a worldwide epidemic. So many people got sick that temporary hospitals such as this one had to be set up to hold them all.

Kurzweil and Joy added that the virus could be far more destructive than a bomb. While an atom bomb dropped on an American city could kill as many as 1 million people, the release of a deadly biological virus could kill tens of millions. The two writers emphasized the need for strict limits on the availability of this type of information. In response to this fear, Congress passed the National Defense Authorization Act in 2016, which requires the federal government to develop a plan for biodefense.

It is unclear how likely it is that someone will create deadly organisms and create intentional public health havoc, but there are other dangers to consider. The fact remains that synthetic genes are composed of hundreds of thousands of A-T and G-C base pairs. People input the information into computers and wait for machines to print out the finished DNA sequences. Just as the spell-check and grammar-check features on word processors depend on humans to program correct spelling and grammar into computers, biological fabrication machines—biofabs for short—depend on humans, too. Humans must input the DNA sequences, and humans develop the machines that print them out. Mistakes are common. In fact, to minimize error, companies often make a series of shorter sequences and then splice them together.

Another concern is that even if mistakes are eliminated entirely, living organisms—even synthetic ones—are still unpredictable. When they leave the controlled world of the laboratory, they interact with the environment in ways that can be unexpected. The fear exists, even among the researchers most closely involved with the technology, that synthetic genes might cause unforeseen public health disasters if they were to somehow get out of the lab.

Biohacking

While professional scientists meddle with the building blocks of life, a group of serious amateurs participate in do-it-yourself biology, or DIYbio. Since CRISPR kits are relatively cheap and widely available, synbio has now become a hobby. Groups of DIYbiologists meet to extract DNA and perform DNA experiments using common household products and tools. While bringing this science into the public realm has some benefits, it also worries some scientists. Although the term biohacking sounds illegal, serious scientists use it to refer to the work of amateurs devoted to arranging cells and genes in creative ways. An article in the *Tech*, MIT's campus newspaper, explained how synthetic biology makes biohacking easier: "The dream … is that biology novices could browse a catalog of ready-made biological parts and use them to create customized organisms."[45] However, experts fear that biohackers might create deadly microbes or alter their own genes in a way that is harmful to them. In addition to fearing for people's safety, scientists worry that since the technology is so new, any bad press will make people fear it even more and set the research field back, limiting its potential to be helpful.

Tom Knight, formerly a senior researcher at MIT, cofounded a synthetic biology company he named Ginkgo Bioworks. He sees value in the ingenuity of DIYbiologists. In fact, he and a group of synthetic biologists at MIT identify themselves as biohackers. However, Knight is also unsure about the safety of making biological information so readily accessible to people who are unsure of how the technology works. He said, "I think if the safety issues can be addressed, there is a big opportunity … It's a huge issue; how do you regulate so [people] don't cause havoc."[46] According to the *Tech*, Ginkgo acknowledged the excitement of amateur scientists by publishing a comic book-style guide to biohacking. Each frame demonstrates a small bioengineering experiment. The company compared backyard genetic engineering to the early days of computer technology, when young entrepreneurs such as Bill Gates of Microsoft launched companies in their basements and garages. In 2018, Ginkgo Bioworks partnered with Bayer (formerly Monsanto) to create a company called Joyn Bio, which will work mainly on using genetic engineering to create more sustainable, environmentally friendly farming methods.

Ginkgo Bioworks encourages amateur biologists to follow their passion and offers resources and internships to those looking to learn more about biotechnology.

A New Kind of Evolution

As scientists become more skilled at engineering and synthesizing the microscopic components of life—whether synthetic or natural—the issues surrounding these accomplishments loom large. Questions about who should profit and how genetic engineering should be controlled have yet to be resolved. However, even as entire industries focus their sights on these extreme versions of genetic engineering, an issue of massive proportions troubles many scientists and members of the public. Some imagine a threatening future if and when genetic engineering—especially the genetic engineering of humans—becomes commonplace.

Altering Nature

Some scientists are researching ways to use CRISPR to stop the spread of disease. One group of researchers is exploring the idea of editing mosquito DNA to make future generations of female mosquitos sterile, or unable to produce offspring. Reducing the mosquito population this way would make it harder for them to spread malaria among humans, but it could also have serious consequences for animals that eat mosquitos.

Another project underway as of 2018 is headed by biologist Kevin Esvelt, who is working to genetically engineer mice in a way that would make them unable to pass Lyme disease to ticks. This would stop ticks from giving the disease to humans. However, as an experienced biologist, Esvelt is aware that editing the mice's genes might have negative outcomes no one has thought of. For this reason, all his tests are being carried out on an uninhabited island so the GE mice can easily be contained if necessary. In an interview, Esvelt said, "I worry every day that I might be missing something profound about the consequences of what we're developing."[1] Many biohackers do not display this type of caution, which makes experts worry about what these amateur scientists might accidentally do.

1. Quoted in "Gene Editing: Last Week Tonight with John Oliver (HBO)," YouTube video, 12:50, posted by Last Week Tonight, July 1, 2018, [illegible].

Some people, including Lee Silver, believe humans could be headed toward a world inhabited by two very different human species. In his book *Challenging Nature*, Silver painted a picture of what the future might hold if genetic engineering proceeds unrestrained. He described two human species, which he called "naturals" and "enhanced," that could result. Naturals will depend on old-fashioned genetic luck. Enhanced humans will have advantages given to them by genes either selected or engineered to provide superior intelligence, talents, looks, and other traits. Each successive generation will have the opportunity to enhance even further the genetic fortunes of their children. Genetic advantages will pass from one generation to another. Eventually, the two groups will differ so greatly they will have evolved into different species.

Silver is particularly troubled by how that imagined scenario would play out in society. When he was interviewed by the BBC, he told reporters,

> *I don't think those people [the naturals and the enhanced] would be able to interact very well and so they will stay apart from each other socially, and ultimately they won't be able to breed with each other ... I think it's going to be a disaster because one group of people who is a different species to the other group of people will no longer have the desire or need to treat that second group with dignity and respect.*[47]

This scenario is not as far-fetched as it once may have seemed. When Charles Darwin proposed his theory of evolution in the 1850s, he explained how species evolve to adapt to challenges in the environment. Some modern biologists believe that genetic engineering signals the end of Darwinian evolution. Humans are now on the brink of directing evolution on their own terms.

Someday—perhaps sooner than people think—humans could hold the power to manipulate genetics as they see fit.

The field of genetic engineering has already achieved remarkable goals in the last 20 years alone. Farmers already grow food crops genetically engineered to resist pests or provide more nutrition. Meat can be grown in vats. Manufacturers weave fiber as strong as steel from the GE milk of goats.

Although things similar to these have been imagined in science fiction for decades, the reality is often slightly different. While it is certain that researchers will continue to make extraordinary discoveries, scientists may never achieve some of the things described in movies and books, such as designer babies and pigs that grow human organs. To some, this is a disappointment, but to others, it is a relief.

NOTES

Introduction: The Future Is Here

1. Robert Krulwich and Jad Abumrad, "Minty E. Coli and Other Bioengineering Feats," National Public Radio, April 29, 2008. www.npr.org/templates/story/story.php?storyId=90014997.

Chapter 1: Farming Pharmaceuticals

2. Freeman J. Dyson, *The Sun, The Genome, and the Internet*. New York, NY: Oxford University Press, 1999, p. 17.
3. Elizabeth Svoboda, "Pharm Animals Crank Out Drugs," *Wired*, February 14, 2007. www.wired.com/medtech/health/news/2007/02/72708.
4. Svoboda, "Pharm Animals Crank Out Drugs."
5. Quoted in Mark Henderson, "Mutant Monkeys Bred to Contract Brain Disorder," *London Times*, May 19, 2008. www.timesonline.co.uk/tol/news/uk/science/article3958803.ece.
6. Quoted in Ashley Feinberg, "These Pigs Glow-in-the-Dark to Save Lives," Gizmodo, December 30, 2013. gizmodo.com/these-pigs-glow-in-the-dark-to-save-lives-1491589459.
7. Quoted in "Genetic Engineering of Food Crops to Produce Drugs Raises Concern," The Organic and Non-GMO Report, September 2002. www.nongmoreport.com/Genetic_engineering.php.
8. Quoted in Margot Roosevelt, "Cures on the Cob," *TIME*, May 19, 2003. www.time.com/time/magazine/article/0,9171,452804-1,00.html.

9. "How FDA Regulates Food from Genetically Engineered Plants," U.S. Food and Drug Administration, last updated January 4, 2018. www.fda.gov/Food/IngredientsPackagingLabeling/GEPlants/ucm461831.htm.

10. Kate Siegel and Suzanne Verity, "What You Need to Know About GMOs," WebMD, April 8, 2016. www.webmd.com/food-recipes/features/truth-about-gmos#4.

Chapter 2: Genetically Engineered Foods

11. G. Bruening and J. M. Lyons, "The Case of the FLAVR SAVR Tomato," *California Agriculture*, July 1, 2000. calag.ucanr.edu/Archive/?article=ca.v054n04p6.

12. Quoted in "Understanding Biotech in Africa; Clipping Regulatory Wings; Public Education of Issues; Urban Legends … Fish-Gene Strawberry; Agony Uncle and Anger Management; Organic Myths and Facts," *AgBioWorld*, April 28, 2004. www.agbioworld.org/newsletter_wm/index.php?caseid=archive&newsid=2107.

13. Juli Locquet, "The Price of Convenience: Fishberry Fields, Forever!," The Organic View, May 16, 2010. www.theorganicview.com/other-shows/featured-guests-on-the-organic-view-radio-show/trends/the-price-of-convenience-fishberry-fields-forever/.

14. Ross Keller, "GMO Misconceptions: Tomatoes with Fish Genes Not 'Fishy,'" Genetic Literacy Project, November 10, 2015. geneticliteracyproject.org/2015/11/10/gmo-misconceptions-tomatoes-with-fish-genes-not-fishy/.

15. Siegel and Verity, "What You Need to Know About GMOs."

16. Locquet, "The Price of Convenience."

17. Siegel and Verity, "What You Need to Know About GMOs."

18. David R. Just and Harry M. Kaiser, "A Good Deal for the Environment and the Poor," *Forbes*, September 14, 2016. www.forbes.com/sites/gmoanswers/2016/09/14/gmos-benefit-environment-poor/#54f726242df4.

19. Alice Park, "7 Things You Need to Know About GMO Salmon," *TIME*, November 19, 2015. time.com/4120648/fda-approved-aquabounty-gmo-salmon/.

20. "Bovine Growth Hormone: To Use or Not to Use, That is the Question," *Living on Earth*, September 8, 1995. www.loe.org/shows/segments.html?programID=95-P13-00036&segmentID=1

21. "Bovine Growth Hormone," *Living on Earth*.

22. Ben Macintyre, "Growing Meat in a Petri Dish," *Calgary Herald*, January 14, 2007. www.pressreader.com/canada/calgary-herald/20070115/282119222064297.

23. "Yes, This Is Actual Meat, but No Animal Died for It," PETA, March 21, 2017. www.peta.org/living/food/memphis-meats-debuts-lab-grown-chicken-clean-meat/.

24. "Yes, This Is Actual Meat," PETA.

25. Just and Kaiser, "A Good Deal."

26. "Three Years Later: Genetically Engineered Corn and the Monarch Butterfly Controversy," Pew Initiative on Food and Biotechnology, 2002, p. 3. www.pewtrusts.org/-/media/legacy/uploadedfiles/wwwpewtrustsorg/reports/food_and_biotechnology/vfbiotechmonarchpdf.pdf.

27. Quoted in Andrew Pollack, "In Lean Times, Biotech Grains Are Less Taboo," *New York Times*, April 21, 2008, p. A16.

Chapter 3: Fabric and Fuel

28. Sarah Kramer, "Genetically Engineered Spider Goats Could Be Biotechnology's Next Big Hit," *Business Insider*, April 28, 2016. www.businessinsider.com/genetic-engineering-goats-spiders-2016-4.

29. Kramer, "Genetically Engineered Spider Goats."

30. Marc Bain, "Your Organic Cotton T-shirt Might Be Worse for the Environment Than Regular Cotton," Quartz, May 28,

2017. qz.com/990178/your-organic-cotton-t-shirt-might-be-worse-for-the-environment-than-regular-cotton/.

31. Yvon Chouinard, "What Does a Clothing Company Know About Genetic Engineering?," Patagonia, Winter 2002. www.patagonia.com/web/eu/patagonia.go?assetid=9134.

32. Andrew Pollack, "Redesigning Crops to Harvest Fuel," *New York Times*, September 8, 2006. www.nytimes.com/2006/09/08/business/08crop.html?fta=y.

33. Freeman Dyson, "Our Biotech Future," *The New York Review of Books*, July 19, 2007. www.nybooks.com/articles/2007/07/19/our-biotech-future/.

Chapter 4: Designing Babies

34. Michael D. Lemonick, "Designer Babies," *TIME*, January 11, 1999, p. 64.

35. Mike Polyakov, "The Ethics of Designer Children," Institute for Ethics and Emerging Technologies, April 10, 2008. ieet.org/index.php/IEET/more/polyakov20080410.

36. Lisa Ko, "Unwanted Sterilization and Eugenics Programs in the United States," PBS *Independent Lens*, January 29, 2016. www.pbs.org/independentlens/blog/unwanted-sterilization-and-eugenics-programs-in-the-united-states/.

37. Nicholas D. Kristof, "Where Is Thy Sting?," *New York Times*, August 12, 2003. www.nytimes.com/2003/08/12/opinion/where-is-thy-sting.html.

38. Kara Rogers, "Chimera," *Encyclopedia Britannica*, accessed on June 19, 2018. www.britannica.com/science/chimera-genetics.

39. Kelly Servick, "Scientists Grow Bullish on Pig-to-Human Transplants," *Science*, September 22, 2017. www.sciencemag.org/news/2017/09/scientists-grow-bullish-pig-human-transplants.

40. Lee Silver, "Human-Animal Chimeras: From Mythology to Biotechnology," Scientific Blogging, February 15, 2007.

41. Quoted in Nicholas Wade, "Chimeras on the Horizon but Don't Expect Centaurs," *New York Times*, May 3, 2003. www.nytimes.com/2005/05/03/science/03chim.html?pagewanted=1&_r=1&oref=slogin.

Chapter 5: Extreme Genetic Engineering

42. Andrew Pollack, "Researchers Take Step Toward Synthetic Life," *New York Times*, January 25, 2008, p. A17. www.nytimes.com/2008/01/25/science/25genome.html.

43. Quoted in Andrew Pollack, "How Do You Like Your Genes? Biofabs Take Orders," *New York Times*, September 12, 2007. www.nytimes.com/2007/09/12/technology/techspecial/12gene.html.

44. Ray Kurzweil and Bill Joy, "Recipe for Destruction," *New York Times*, October 17, 2005. www.nytimes.com/2005/10/17/opinion/17kurzweiljoy.html?scp=1&sq=recipe%20for%20destruction&st=cse.

45. Carolyn Y. Johnson, "As Synthetic Biology Becomes Affordable, Amateur Labs Thrive," *Tech*, September 16, 2008. tech.mit.edu/V128/N39/biohack.html.

46. Quoted in Johnson, "As Synthetic Biology Becomes Affordable, Amateur Labs Thrive."

47. Quoted in "Back to the Future," BBC News, December 1999. news.bbc.co.uk/hi/english/static/special_report/1999/12/99/back_to_the_future/lee_silver.stm.

DISCUSSION QUESTIONS

Chapter 1: Farming Pharmaceuticals

1. In what ways has genetic engineering changed the traditional use of medicinal plants?
2. How do both ordinary people and biotechnology companies benefit from pharmaceuticals developed by genetic engineering?
3. What are the pros and cons of genetically engineering animals for the sake of scientific research?

Chapter 2: Genetically Engineered Foods

1. Describe the main conflict between supporters and critics of Roundup Ready soy and Bt corn.
2. Would you eat meat grown in a vat? Why or why not?
3. What are some situations that might decrease people's resistance to genetically engineered foods?

Chapter 3: Fabric and Fuel

1. Can you think of ways to use spider silk aside from the ones mentioned in this chapter?
2. Which do you think is better—organic cotton or conventional cotton? Explain your answer.
3. What are some issues surrounding biofuel?

Chapter 4: Designing Babies

1. Why is eugenics dangerous?
2. Why are some scientists concerned about the method used in gene therapy to carry replacement genes into the cells?
3. Why might people object to genetic engineering that results in organisms that have obvious human features, while they do not object to organisms that do not appear human?

Chapter 5: Extreme Genetic Engineering

1. Why does synthetic biology lead people to question the meaning of life itself?
2. How is biohacking similar to and different from the early days of computer programming?
3. Do you agree with Lee Silver's predictions about the future? Why or why not? How might we stop the inequality he foresees?

ORGANIZATIONS TO CONTACT

Biotechnology Innovation Organization (BIO)
1201 Maryland Ave. SW, Ste. 900
Washington, DC 20024
(202) 962-9200
info@bio.org
www.bio.org

BIO is the world's largest biotechnology organization. It advocates the research, development, and commercial distribution of products that result from genetic engineering.

Center for Genetics and Society
1122 University Ave #100
Berkeley, CA 94702
(510) 665-7760
www.geneticsandsociety.org

This nonprofit information and public affairs organization encourages responsible use and effective social control of new human genetic and reproductive technologies. It supports some medical applications and opposes others that threaten to divide human society.

Council for Responsible Genetics
660 Pennsylvania Ave. SE
Washington, DC 20003
(202) 547-9359 ext. 24
genewatch@icta.org
www.councilforresponsiblegenetics.org

This nonprofit organization encourages public debate about the social, ethical, and environmental impacts of genetic technology.

Greenpeace USA
702 H St. NW, Ste. 300
Washington, DC 20001
(202) 462-1177
info@wdc.greenpeace.org
www.greenpeace.org/usa

Greenpeace lobbies the government to limit or ban products that are the result of genetic engineering. The organization has offices worldwide.

Organic Consumers Association (OCA)
6771 S. Silver Hill Dr.
Finland, MN 55603
(218) 226-4164
www.organicconsumers.org

The OCA is an online, grassroots public interest organization that deals with food safety, industrial agriculture, and genetic engineering. Its focus is on promoting the views and interests of consumers of organic products.

Union of Concerned Scientists (UCS)
2 Brattle Sq.
Cambridge, MA 02138
(617) 547-5552
www.ucsusa.org

This science-based nonprofit organization that works for a healthy environment and a safer world began as a collaboration of students and faculty members at MIT. UCS combines independent scientific research and citizen action to promote responsible government and corporate policies for cleaner and healthier air, energy, food, and transportation.

FOR MORE INFORMATION

Books

Gray, Leon. *Genetic Modification: Should Humans Control Nature?* London, UK: Franklin Watts, 2015.

This book investigates the moral implications of genetic engineering, including opinions from many sides of the issue.

Henneberg, Susan. *Cloning*. Costa Mesa, CA: Saddleback Educational Publishing, 2015.

Cloning has been featured in science fiction for decades, but the reality—discussed in this book—often looks far different.

Lachner, Elizabeth. *Bioengineering*. New York, NY: Britannica Educational Publishing, 2016.

Lachner's book offers an overview of bioengineering, where biological and technological sciences work together. It includes information about synthetic biology and genetic engineering.

Rooney, Anne. *Genetic Engineering and Developments in Biotechnology*. St. Catharines, ON: Crabtree Publishing, 2017.

Rooney's readers take a look at biotechnology, the job of a genetic engineer, and the genetic engineering process.

Smundak, Katharina. *The Truth Behind GMOs*. New York, NY: Rosen Central, 2018.

With so much contradictory information available about GE crops, it can be difficult for consumers to know what to think. This book discusses the science behind GMOs to give people a clearer picture of the issue.

Websites

Alliance for Humane Biotechnology

www.humanebiotech.org

This nonprofit group is committed to informing the public about the social impacts of genetic engineering and other biotech advances. Its website provides information about topics such as synthetic biology, genetic privacy, and designer babies.

As You Sow

www.asyousow.org

This nonprofit organization promotes corporate responsibility toward the environment and society. The organization publishes a blog about these issues, including those related to the genetic modification of crops and animals for profit.

GeneWatch UK

www.genewatch.org

This nonprofit group monitors the development of genetic technology and makes sure people are well-informed and able to voice their opinions about how genetic engineering should be used. Its website provides a more global perspective on genetic modification.

New Harvest

www.new-harvest.org

The mission of New Harvest is to advance research into animal product substitutes. The website includes links to articles about cultured meat substitutes.

Northwest Resistance Against Genetic Engineering

nwrage.org

This organization seeks to ban genetic engineering. The website includes a complete list of resources, articles, and organizations that lobby against genetic engineering.

INDEX

N

O

P

R

S

PICTURE CREDITS

Cover Westend61/Brand X Pictures/Getty Images; p. 7 Houghton Library, Harvard University, Cambridge, Mass.; p. 10 PA Images via Getty Images; p. 12 dianaduda/Shutterstock.com; p. 15 Moonborne/Shutterstock.com; p. 17 wunchana/Shutterstock.com; p. 19 Andrei Dubadzel/Shutterstock.com; p. 21 Kimberly Boyles/Shutterstock.com; p. 24 bullet74/Shutterstock.com; p. 27 Parkpoom Chulapoomphinit/Shutterstock.com; p. 34 Steve Bly/Photographer's Choice RF/Getty Images; p. 40 Tatchaphol/Shutterstock.com; p. 43 Jill Lang/Shutterstock.com; p. 48 Lindsay7/Shutterstock.com; p. 50 fuyu liu/Shutterstock.com; p. 51 Alf Manciagli/Shutterstock.com; p. 54 busypix/E+/Getty Images; p. 59 Monkey Business Images/Shutterstock.com; p. 64 vchal/Shutterstock.com; p. 67 Olaf Kruger/Getty Images; p. 72 Abeselom Zerit/Shutterstock.com; p. 76 Nataliya Arzamasova/Shutterstock.com; p. 79 chromatos/Shutterstock.com; p. 80 Underwood Archives/Getty Images; p. 83 John Tlumacki/The Boston Globe via Getty Images; p. 85 Eetu Mustonen/Shutterstock.com.

ABOUT THE AUTHOR

Michelle Denton received her bachelor's degree in English and creative writing from Canisius College in 2016, graduating cum laude from the All-College Honors Program. She lives in Buffalo, New York, and has made writing her full-time career.

KATE MCKINNON